DIELS-ALDER REACTIONS

DIELS-ALDER REACTIONS

ORGANIC BACKGROUND
AND PHYSICO-CHEMICAL ASPECTS

by

A. WASSERMANN
Reader in Chemistry
University College, London

ELSEVIER PUBLISHING COMPANY
AMSTERDAM – LONDON – NEW YORK
1965

ELSEVIER PUBLISHING COMPANY
335 JAN VAN GALENSTRAAT, P.O. BOX 211, AMSTERDAM

AMERICAN ELSEVIER PUBLISHING COMPANY, INC.
52 VANDERBILT AVENUE, NEW YORK, N.Y. 10017

ELSEVIER PUBLISHING COMPANY LIMITED
RIPPLESIDE COMMERCIAL ESTATE BARKING, ESSEX

LIBRARY OF CONGRESS CATALOG CARD NUMBER 65-13901
WITH 9 ILLUSTRATIONS AND 24 TABLES

ALL RIGHTS RESERVED
THIS BOOK OR ANY PART THEREOF MAY NOT BE REPRODUCED
IN ANY FORM,
INCLUDING PHOTOSTATIC OR MICROFILM FORM,
WITHOUT WRITTEN PERMISSION FROM THE PUBLISHERS

PRINTED IN THE NETHERLANDS

PREFACE

Cyclo-additions, $mA + nB \rightarrow C$ ($m \geq 1$; $n \geq 1$; $A \neq B$ or $A = B$), where the cyclic product C contains σ-bonds formed at the expense of π-bonds in there actants A and B, can produce four-, five-, six-, seven- and eight-membered rings. Examples are the dimerisation of alkenes, the addition of so-called 1,3-dipoles to dipolarophiles, Diels–Alder associations, the reactions of ozone with an anthracene derivative and the conversion of acetylene into cyclooctatetraene. These thermal or photochemical associations and related reactions, *e.g.*, the conversion of butadiene into cyclobutene or the cyclisation of hexatriene, are important and their physico-chemical aspects are interesting. After this monograph was completed a theoretical interpretation of the stereospecificity of certain inter- and intra-molecular cyclo-additions was published and it appears that there is a new field opening up for the application of certain aspects of molecular orbital calculations. For these and other reasons it would be worth while to write a book on cyclo-additions in general, but I have not dared to undertake this rather ambitious task. Instead I discuss here Diels–Alder associations only, but I hope that some of the physicochemical methods which are described will be useful to workers in the field of other cyclo-additions. Previous review articles on Diels–Alder reactions or chapters in books have treated preparative, stereochemical and mechanistic aspects, but a summarising paper or monograph, containing the organic background together with a rather detailed account of physicochemical measurements which are useful for establishing a successful mechanistic theory has not yet been published. I decided, therefore, to fill this gap.

The book is written mainly for research workers interested in correlations between structure and mechanism, and who would find it useful, as a side line, to employ Diels–Alder reactions as "tools" for the development of transition state theory. This latter aspect is briefly indicated in the introduction to Chapter 3 and in Section 4,1. Undergraduate students, on the other hand, who have not yet acquired a good knowledge of reaction kinetics and its statistical mechanical background should consult other texts before reading Chapters 3-5 because no derivations are given of fundamental formulae frequently mentioned.

No attempt is made to cover the literature on Diels–Alder reactions completely or to discuss papers before the first classical publication of Diels and Alder in 1928. The topics to be described have been selected according to personal taste. If experiments to clear up controversial points in Chapter 5 are stimulated by this book, its purpose will be fulfilled.

Many thanks are due to Sir Christopher Ingold and to Lady Ingold for inspiration and help while doing research on Diels–Alder reactions and while completing this booklet. I am very grateful to Dr. A. G. Davies for having written Chapter 1 and for comments on the other chapters; and to Professor C. Eaborn and Dr. D. V. Banthorpe for constructive criticism.

London, July 1965. A.W.

CONTENTS

Preface v

Chapter 1. ORGANIC BACKGROUND 1
  1.1 General 1
  1.2 Acyclic dienes 3
  1.3 Vinylcycloalkenes and vinylarenes 7
  1.4 One-ring dienes 9
  1.5 Arenes 13
  1.6 Dimethylenecycloalkanes 15
  1.7 Polyenes 18
  1.8 Cycloalkenes and cycloalkynes as dienophiles 20

Chapter 2. STEREOCHEMISTRY AND RELATED TOPICS . . . 24
  2.1 Conformation of diene and retention of stereochemical integrity of dienophile 24
  2.2 Isomeric Diels–Alder adducts 26
  2.3 Trimers and polymers 31
  2.4 Isomerisation of Diels–Alder adducts 34

Chapter 3. EQUILIBRIA 39
  3.1 Theoretical relationships and results 40
  3.2 Heats of reaction 43
  3.3 Entropies of reaction 44
  3.4 Equilibria in solution 45
  3.5 Mode of alignment of diene and dienophile 46

Chapter 4. KINETICS AND CATALYSIS 48
  4.1 Associations in the gaseous and condensed state 48
  4.2 Formation of geometrically isomeric adducts 58
  4.3 Diels–Alder retrogressions 61
  4.4 Added substances without influence on the rate of Diels–Alder reactions 64
  4.5 Catalysis of thermal Diels–Alder reactions 65
  4.6 Photo-associations 71

Chapter 5. MECHANISM 74
  5.1 One-step mechanism; activation energy 74

5.2	One-step mechanism: entropy of activation	92
5.3	Two-step mechanism of uncatalysed thermal Diels-Alder reactions	93
5.4	Diels–Alder retrogressions and isomerisation of endo-exo-adducts	96
5.5	Catalysed associations	96

Appendix 99

References 101

Subject Index 110

Chapter 1

ORGANIC BACKGROUND

The extensive literature describing the versatile and useful synthetic principle of 1,4-additions to conjugated multiple bonds, known as the Diels–Alder reaction, has been surveyed in many reviews, some of which are listed in ref. 1–4. This chapter deals with the salient features of such associations, thereby giving the organic background of the subsequent physico-chemical discussions.

1. 1 General

In Diels–Alder associations, a diene adds to an olefinic or acetylenic dienophile, as shown, to give an adduct with a six-membered ring which may be bridged.

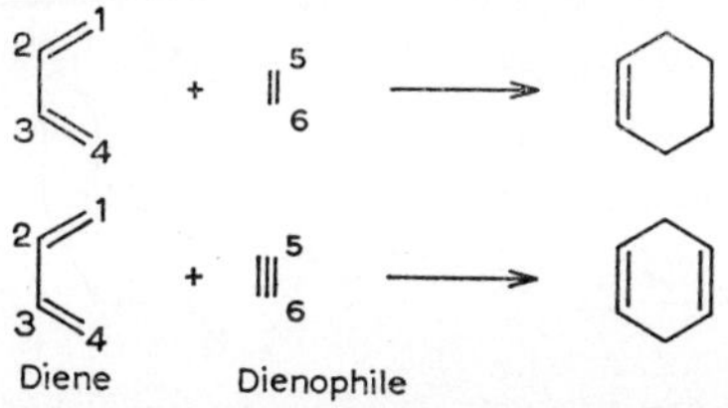

Diene   Dienophile

The six atoms involved in forming the ring are usually all carbon atoms, that is the dienophiles contain olefinic or acetylenic bonds, and only reactions of this type will be considered here. Examples are also known[1], however, in which the diene contains a conjugated double and triple bond, $C=C-C\equiv C$, and others in which the ring closure may take place at atoms other than carbon. Thus the "diene" component may contain the structure $C=C-C=O$, $C=C-C=N$, or $N=C-C=N$, and the dienophile may consist of the groupings C=O, N=O, C≡N, C=N, N=N, O=O, and S=N. Reactions of this kind, to give heterocyclic products, are reviewed in ref. 2. The structure of the diene and dienophile may vary widely, and sections 2–7 of this chapter are divided on the basis of the structure of the diene, while section 8 deals with reactions involving some special dienophiles not mentioned in the earlier sections.

Stereochemical, thermodynamic, kinetic and mechanistic aspects are dealt with in the following chapters. Anticipating more detailed discussions it is pointed out here that many dienes can exist in a *cisoid* and *transoid* conformation*,

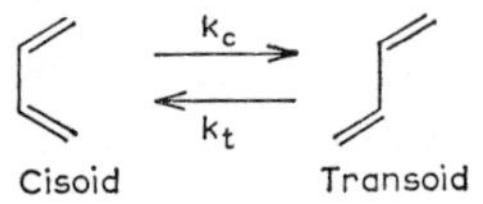

$$\text{Cisoid} \underset{k_t}{\overset{k_c}{\rightleftarrows}} \text{Transoid} \tag{3}$$

where k_c and k_t are velocity coefficients. The transoid species must be converted into the cisoid form, before Diels–Alder addition products are formed. Given the cisoid conformation of the diene, there remains the possibility that substituents originating in the dienophile might be oriented towards (endo) or away from (exo) the newly formed double bond in the erstwhile diene. These

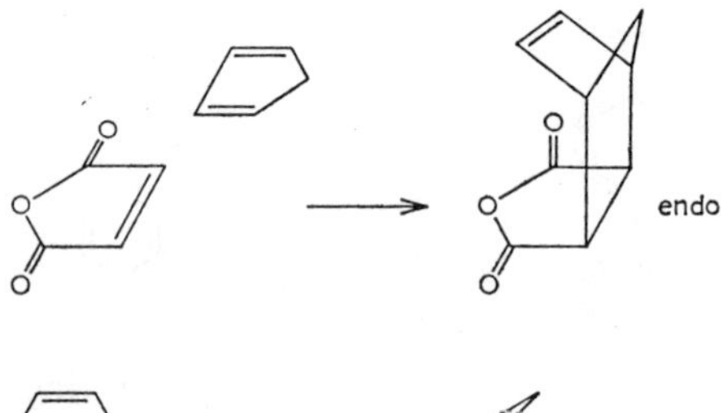

$$\longrightarrow \quad \text{endo} \tag{4}$$

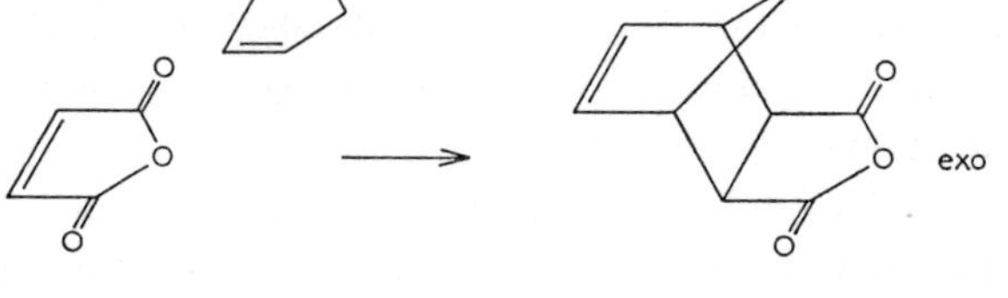

$$\longrightarrow \quad \text{exo} \tag{5}$$

---

* Cisoid and transoid species ressemble cis and trans isomers, except that geometrical restrictions are less rigid.

situations are illustrated in (4) and (5) for the addition of cyclopentadiene to maleic anhydride.

Some Diels–Alder reactions are exothermic and reversible as shown in Chap. 3. The rate of these associations is relatively large if the dienophile contains polar substituents, *e.g.*, carbonyl. For preparative purposes thermal Diels–Alder reactions can be carried out with or without solvent within a large temperature range, *e.g.*, 0–200°. A few catalyzed and photochemical Diels–Alder reactions have been investigated (see Chap. 4) but the possibilities of these techniques are not yet fully exploited.

1. 2 Acyclic dienes

Acyclic 1,3-dienes are excellent reagents for dienophiles, and often give the adducts in almost quantitative yield. For example, butadiene itself reacts quantitatively with maleic anhydride in benzene at 100° in 5 h, or at room temperature in 12 h, forming *cis*-1,2,5,6-tetrahydrophthalic anhydride (I)[5-7]

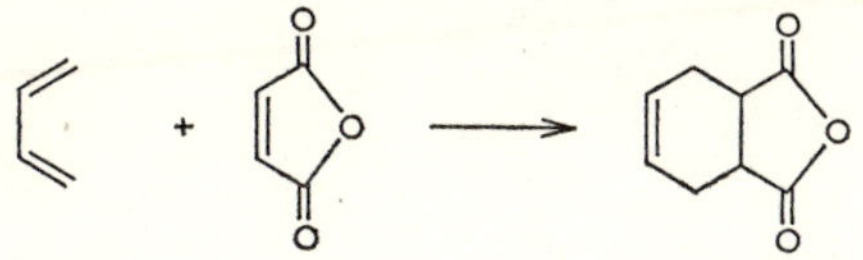

The conditions under which other typical dienophiles react with butadiene are summarised in Table 1. *αβ*-Unsaturated aldehydes (*e.g.* acrolein, crotonaldehyde, and cinnamaldehyde), ketones (*e.g.* methyl vinyl ketone and benzalacetophenone), and acids (*e.g.* acrylic acid, crotonic acid, and cinnamic acid) and their esters react readily. *p*-Benzoquinones can be caused to react with either one or two moles of the diene to give tetrahydronaphthoquinones (6) or octahydroanthraquinone according to reactions (6) or (7)[8].

TABLE 1
REACTIONS OF DIENOPHILES WITH BUTADIENE

Dienophile	Solvent[a]	Temperature	Time[b] (h)	Yield[b] %	Ref.
Maleic anhydride	Benzene	100°	5	100	5–7
$CH_2=CH-CH=O$	0	100°	1	100	5
$CH_2=CH-CO-CH_3$	0	140°	8–10	75–80	14
$CH_2=CEt.NO_2$	0	120–130°	4	37	15
1,4-Benzoquinone	Benzene	Room temp.	Several days	—	16
$CH_2=CH.O.CO.Me$	0	180°	12	6	11
$Me.CO.CH=CH.CO.Me$	Ethanol	100°	12	—	17
$HO_2C.C\equiv C.CO_2H$	Dioxan	170–180°	—	34	13,18
$(EtO_2C)_2C=C(CO_2Et)_2$	0	170°	8	75	19
$Me-CH=CH-CO_2H$	0	150–200°	6	—	20

Notes [a] 0; no solvent used.
[b] —; is not quoted.

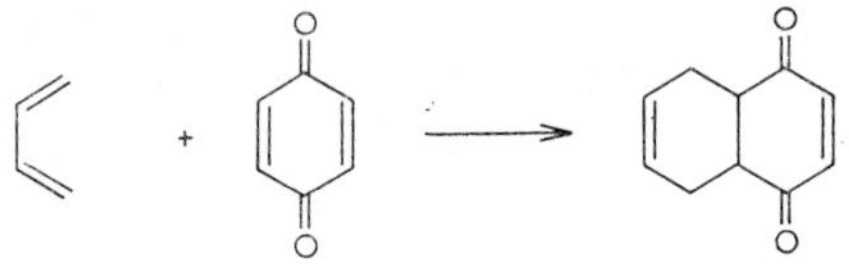

(6)

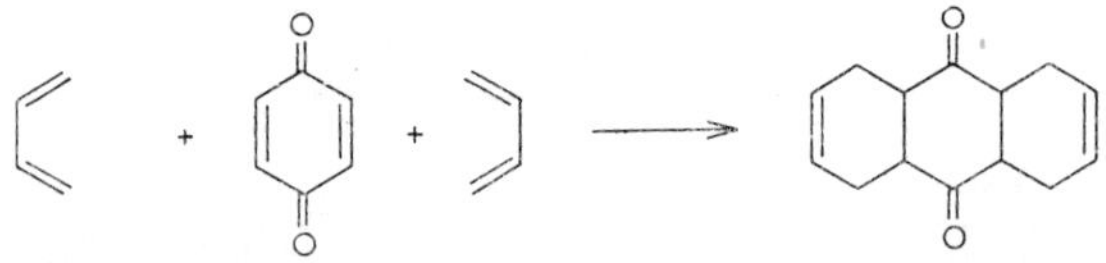

(7)

Acrylonitrile forms the adduct during several days at room temperature; fumaronitrile needs a longer period at high temperatures, but tetracyanoethylene is one of the most reactive dienophiles and gives the adduct[9] in almost quantitative yield below room temperature, thereby providing a method for analysis of 1,3-dienes[10].

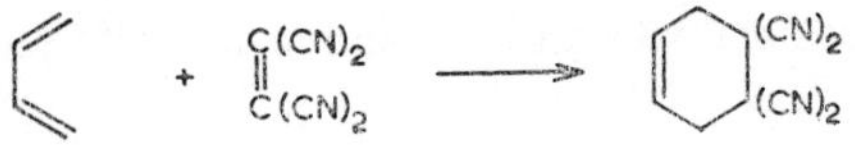

Ethylenic compounds such as allyl alcohol and its esters and halides, and vinyl halides or esters, are relatively unreactive dienophiles, but can frequently be induced to react under forcing conditions, as shown in (8)[11] and (9)[12]; yields of products are given under the formulae

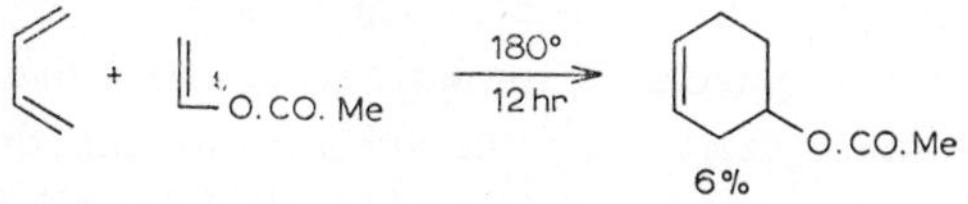

$$\text{butadiene} + \text{CH}_2=\text{CH-O.CO.Me} \xrightarrow[12\text{hr}]{180°} \text{cyclohexenyl-O.CO.Me} \quad 6\% \tag{8}$$

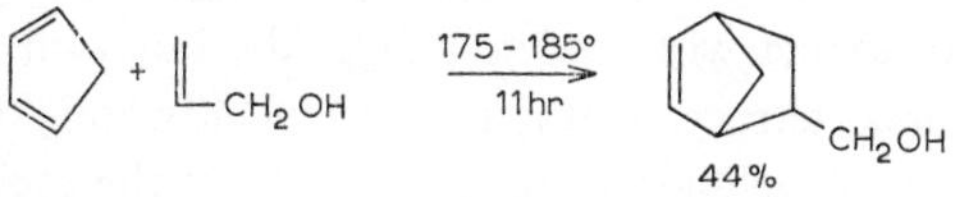

$$\text{cyclopentadiene} + \text{CH}_2=\text{CH-CH}_2\text{OH} \xrightarrow[11\text{hr}]{175-185°} \text{norbornenyl-CH}_2\text{OH} \quad 44\% \tag{9}$$

In the acetylene series, acetylene dicarboxylic acid and its esters have been used most frequently as dienophiles. The reaction occurs readily with all types of dienes; that with butadiene itself[13] is illustrated by association (10).

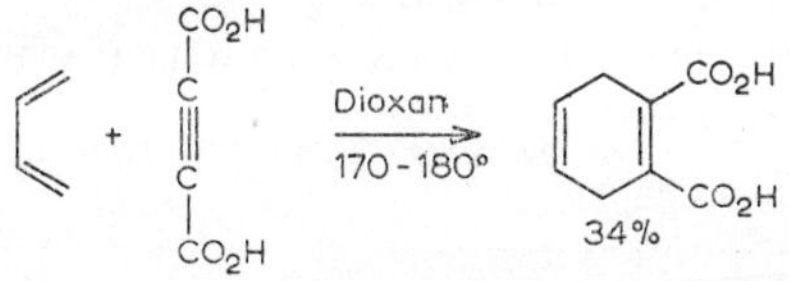

$$\text{butadiene} + \text{HO}_2\text{C-C≡C-CO}_2\text{H} \xrightarrow[170-180°]{\text{Dioxan}} \text{1,2-dihydrophthalic acid} \quad 34\% \tag{10}$$

Many further examples of dienophiles are given in succeeding sections. We return now to a consideration of the diene component of the reaction.

cis-Alkyl or aryl substituents in the 1-position of the diene reduce its reactivity by sterically hindering the formation of the cisoid conformation. For example, *cis*-1-phenylbuta-1,3-diene (IIa and IIb) gives only a 4.3–5.2% yield of the adduct with maleic anhydride under the conditions in which the *trans*-compound (III) reacts quantitatively[21].

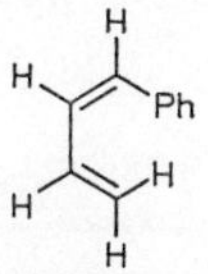

Cisoid cis (IIa)   Transoid cis (IIb)   Cisoid trans (III)

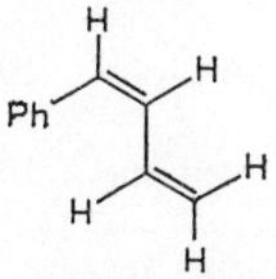

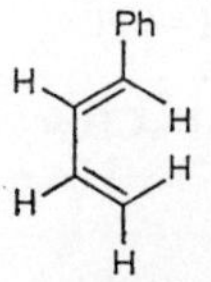

Bulky substituents in the 2,3-positions similarly interfere sterically in the cisoid conformation, and reduce the reactivity of the diene. Thus, whereas 2,3-dimethylbuta-1,3-diene shows a normal reactivity[8], 2,3-diphenylbuta-1,3-diene is somewhat less reactive[22] and 2,3-di-*tert*-butylbuta-1,3-diene[23] (in contrast to the isomeric 1,3-disubstituted compound)[24] is completely inert.

1,1-Disubstituted butadienes similarly react with difficulty, and the steric hindrance exerted by the substituents can divert the reaction into a second path. 1,1-Dimethylbuta-1,3-diene does not form a normal adduct with maleic anhydride, but with the very reactive dienophile tetracyanoethylene, it reacts overnight to give 69% of the normal cyclohexene (IV) and 11% of the cyclobutane derivative (V)[25].

Cis-substituents in both the 1- and the 4-positions, as in *cis,cis*-1,4-diphenylbuta-1,3-diene, render the diene inert.

TABLE 2
REACTIONS OF ACYCLIC 1,3-DIENES WITH MALEIC ANHYDRIDE

Diene	Solvent[a]	Temp.	Time (h)	Yield (%)	Ref.
$CH_2=CH-CH=CH_2$	Benzene	100°	5	100 }	5–6
		20°	12	100 }	
$CH_2=CMe-CH=CH_2$	Benzene	100°	Several	100	26
$CH_2=C(C_8H_{17})-CH=CH_2$	Benzene	Boiling	24	47	27
cis-$MeCH=CH-CH=CH_2$	0	100°		*	28
trans-$MeCH=CH-CH=CH_2$	0	100°	25	99	28
cis-$PhCH=CH-CH=CH_2$	Toluene {	Boiling	4	99	29
		20°	18	71	29
trans-$PhCH=CH-CH=CH_2$	Toluene {	Boiling	4	5	29
		20°	168	0	29
$CH_2=CCl-CH=CH_2$	0	50°	Overnight	77	30
$MeCH=CMe-CMe=CHMe$	Benzene	20°	24	100	7
$CH_2=CPh-CPh=CH_2$	Benzene	Boiling	Overnight	94	22
$C_6H_{13}CH=CH-CH=CH.CO_2H$	Acetone	100°	18	100	31
$HO_2C.CH=CH-CH=CH.CO_2H$	Benzene	100°	18–38	80	7

[a] 0 ; no solvent used.

* slow polymerization.

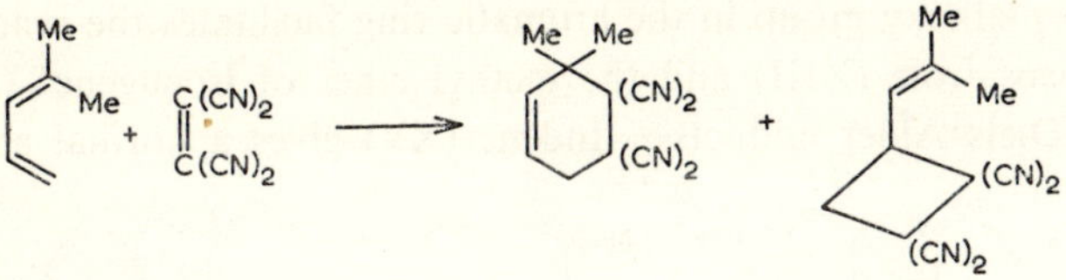

(IV) (V)

The conditions under which various substituted butadienes react with maleic anhydride are summarised in Table 2.

1. 3 Vinylcycloalkenes and vinylarenes

If one of the double bonds of the diene is contained in an alicyclic ring (as in 1-vinylcyclohexene), or if both double bonds are in two different rings (as in 1,1′-bicyclohexenyl), the diene system reacts readily with dienophiles. 3-Methylene-cyclohexenes (VI) or fused-ring systems in which the diene is constrained in a rigid transoid conformation (VII) and (VIII), however, do not take part in Diels–Alder reactions.

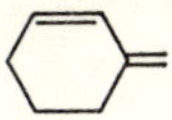

(VI) 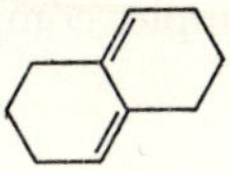   (VII) 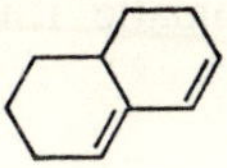   (VIII)

For example, cholesta-7,14-diene (IX) forms a maleic anhydride adduct, but the 6,8(14)-, 7,9(11)-, and 8,14-dienes (X–XII) are inert[32].

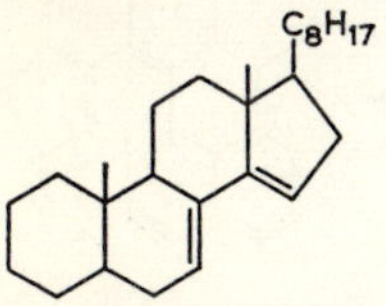

(IX) 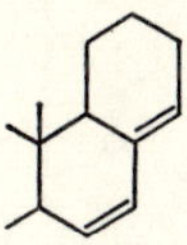   (X) 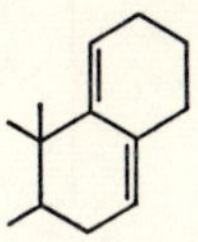   (XI) 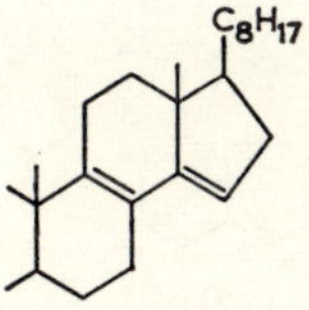   (XII)

Vinylarenes react with dienophiles in a normal manner, even though the addition interrupts the conjugation of the aromatic ring. The reactions are usually carried out in boiling xylene. Styrene and stilbene give a co-polymer with maleic anhydride, but a *p*-alkoxy group in the aromatic ring facilitates the reaction, and isosafrole (XIII) and the methyl ether of isoeugenol (XIV) give Diels–Alder adducts[33]. Indene (XV) gives a normal adduct

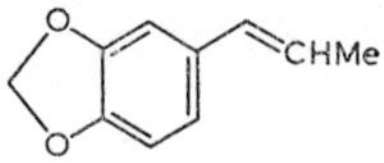

(XIII) 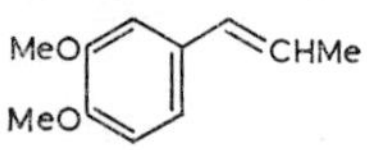   (XIV) 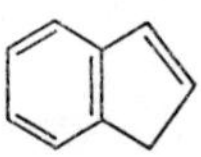   (XV)

under forcing conditions[34], and 1,1-diphenylethylene reacts with maleic anhydride probably in the following manner[35].

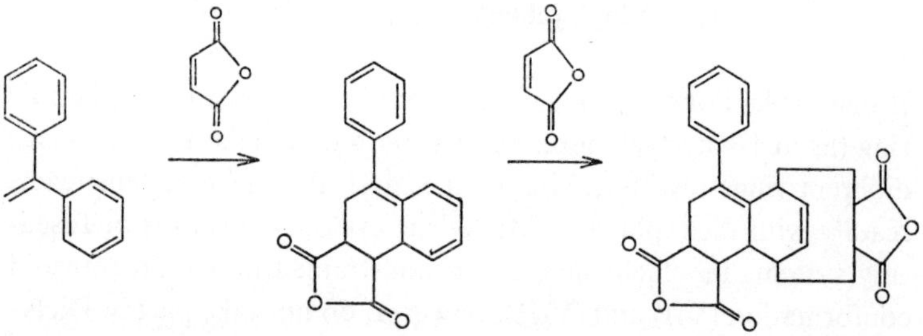

The double bonds of the diene can be contained formally in two different aromatic rings. Simple compounds such as biphenyl

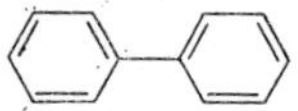

(XVI) 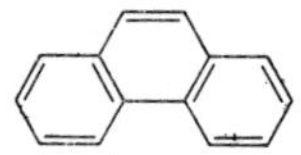   (XVII) 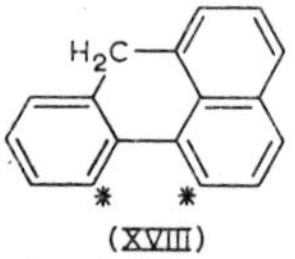   (XVIII)

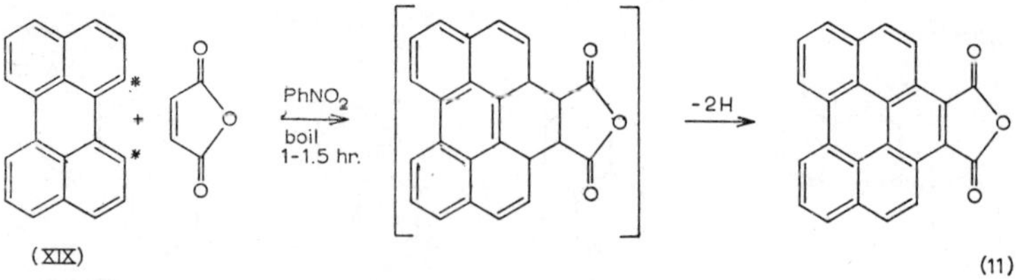

(XIX)

(11)

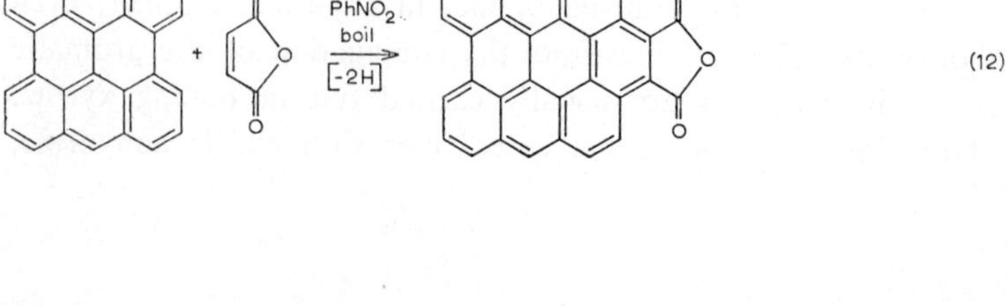

(12)

(XVI) and phenanthrene (XVII) do not react with dienophiles, but benzanthrene (XVIII) and perylene (XIX) do, at the positions marked with an asterisk. Two examples of such additions are shown by reactions (11)[36] and (12)[37].

1. 4 One-ring dienes

The cyclopentadienes are reactive towards dienophiles, the less active dienophiles, such as allyl alcohol, vinyl chloride, and vinyl acetate reacting at elevated temperatures. Even propylene[38] and ethylene[39] will add under forcing conditions.

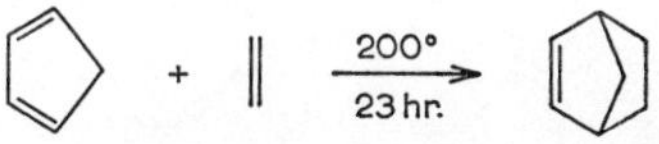

yield 74%.

Cyclopentadiene can act as both diene and dienophile, and readily forms a dimer[40,41].

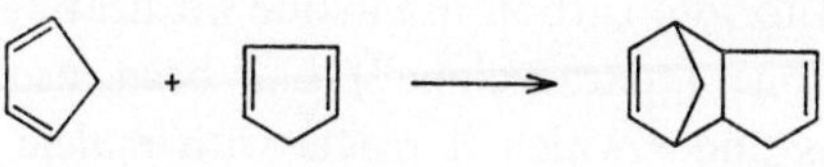

At relatively high temperature, the dimer functions as the dienophile towards a further cyclopentadiene molecule and n-mers (n~6) can be formed by addition of cyclopentadiene to an (n—1)-mer, as shown below for the formation of the trimer[42].

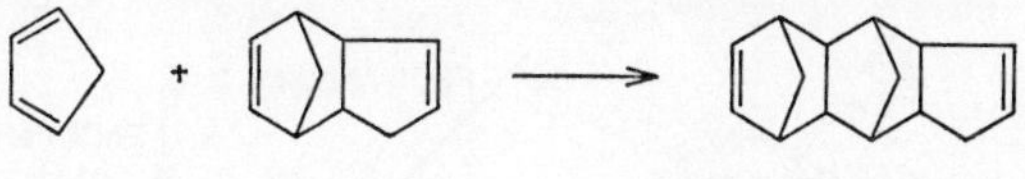

Hexachlorocyclopentadiene forms a maleic anhydride adduct[42], which is used in the manufacture of fire resistant paints. The Diels–Alder reactions of hexachlorocyclopentadiene are also used in the preparation of insecticides, two of which are called Dieldrin and Aldrin:

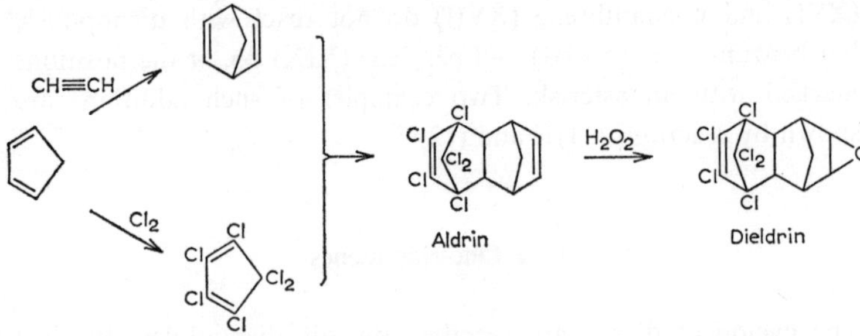

The diene system in the fulvene ring (XX) reacts normally with dienophiles, but the reaction is reversible, and should be conducted at as low a temperature as possible for good yields of the adduct to be obtained[43].

Derivatives of cyclopentadienone (XXI) form adducts similarly, but these tend to lose carbon monoxide on heating. Tetraphenylcyclopentadienone ("tetracyclone") has been used particularly. The conditions under which it reacts with maleic anhydride are shown by reaction (13)[44].

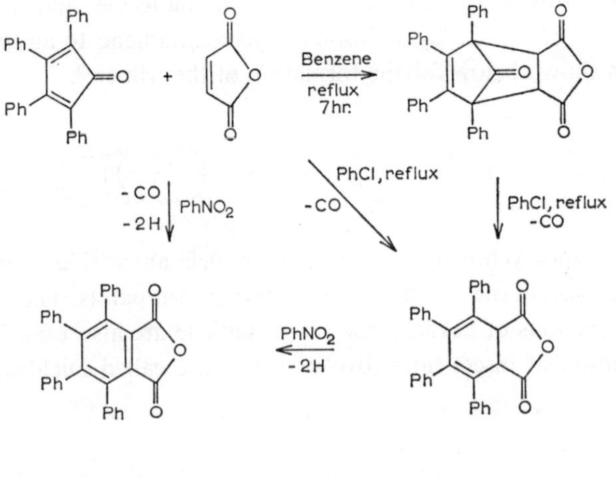

The primary adducts obtained with acetylenic dienophiles are generally not stable, and lose carbon monoxide to give aromatic compounds directly, *e.g.*[45]

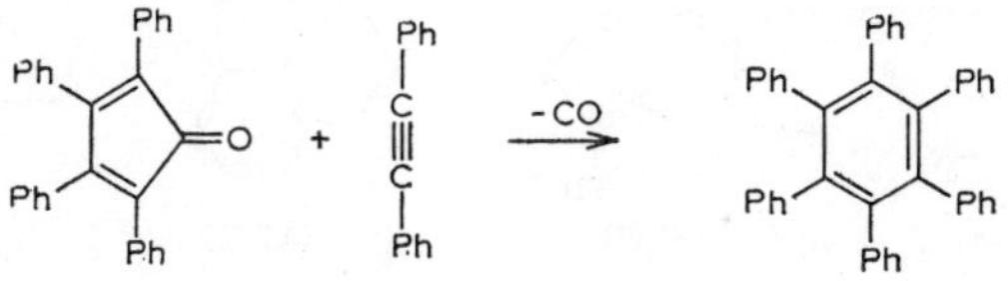

Furan reacts with maleic anhydride in ether on gentle warming

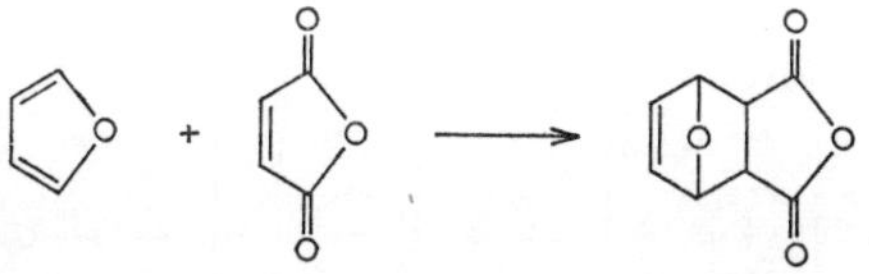

(XXII)

to give the adduct (XXII)[46], in quantitative yield and it reacts with ethylene at 150–155° during 16 hours to give 3,4-epoxycyclohexene[47] (yield 5—8%). Pyrroles usually undergo α-substitution when treated with dienophiles, but *N*-benzylpyrrole with acetylene dicarboxylic acid gives the normal adduct (14)[48], in low yield and both furan and the *N*-methylpyrrole can be used as scavenger for the dienophile benzyne (see later).

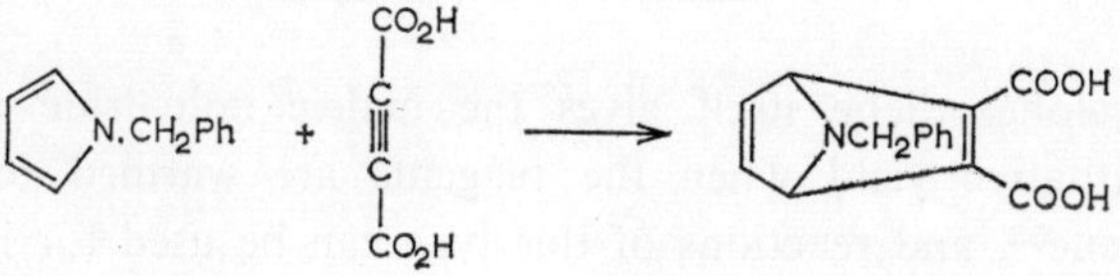

(14)

No Diels–Alder adducts have apparently been formed from thiophen itself, but maleic anhydride adds to the substituted thiophens (XXIII)[49] and (XXIV)[50] (with elimination of H_2S in the latter case), presumably because of the resonance stabilisation of the adducts.

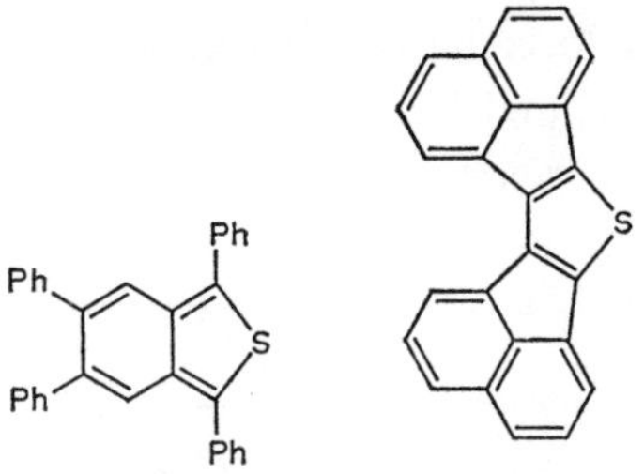

The thiophene *S*-dioxides react with maleic anhydride with the elimination of sulphur dioxide to re-form a diene system, which then adds a second dienophile molecule (15)[51].

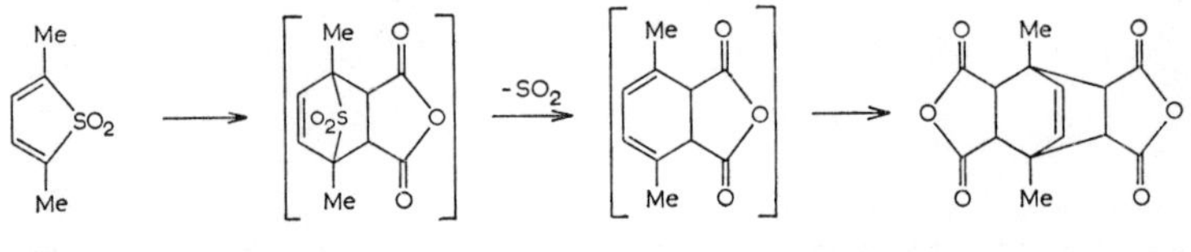

The cyclohexadienes (*e.g.* α-phellandrene, (XXV)[52] and α-terpinene (XXVI)[52] are somewhat less reactive than the cyclopentadienes, but the addition of many dienophiles still takes place in dilute solution at room temperature.

Cyclohexadiene itself gives the maleic anhydride adduct in quantitative yield when the reagents are warmed together in benzene[53], and reactions of this type can be used for identifying the cisoid 1,3-diene system in steroids such as ergosterol[54].

With acetylenic dienophiles, cyclohexadienes give the bicyclo-[2,2,2]octadiene system; if this is heated, it loses an alkene in a reversed Diels–Alder reaction, leaving an aromatic ring. This reaction has been used to determine the structure of some cyclohexadienes, *e.g.*, α-terpinene (XVII)[55].

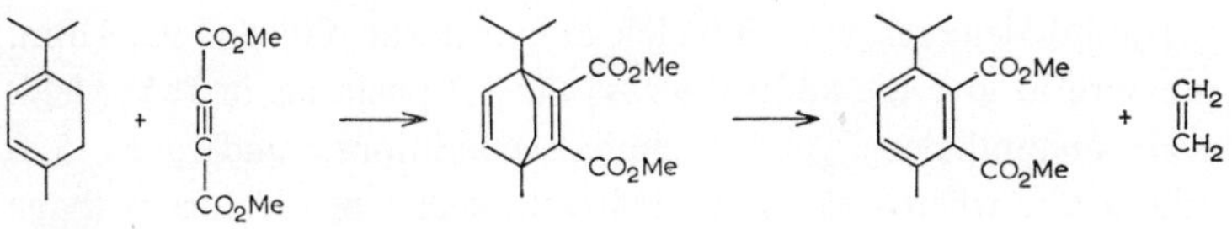

The α-pyrones (XXVIII), like the cyclopentadienes and thiophen dioxides, can liberate a stable molecule—in this case carbon dioxide—from the bicyclic adduct, to give a cyclohexadiene derivative[56].

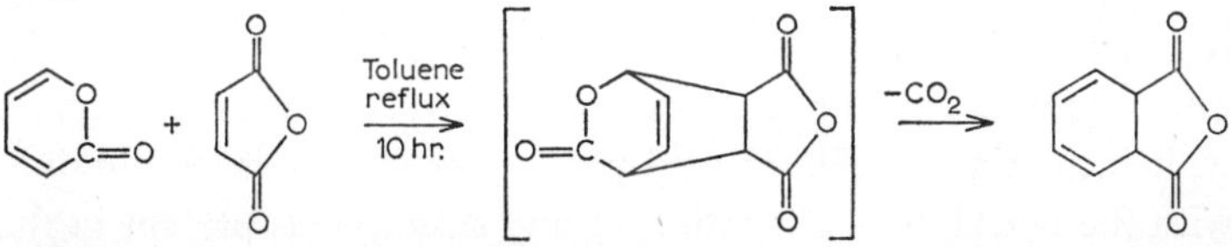

(XXVIII)

Cycloheptadiene reacts normally to give a bicyclo-[3,2,2]-nonane derivative[57]. *cis,trans*-Cycloocta-1,3-diene forms a maleic anhydride adduct[58], but *cis,cis*- and *cis,trans*-cyclodeca-1,3-diene are inert, because of steric strain in forming the cisoid conformation[59].

1. 5 Arenes

Benzene itself does not undergo a thermal Diels–Alder reaction*, but quinol and its monomethyl ether react with maleic anhydride at about 200° to give the 2,5-adduct (XXIX) in low yield[60].

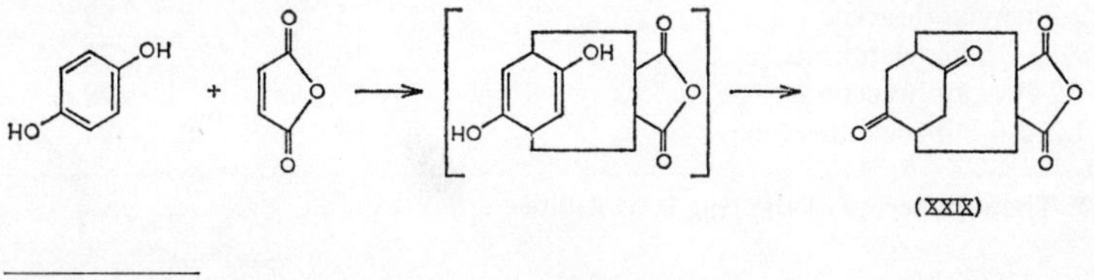

(XXIX)

---

* However, benzene does react with maleic anhydride, under the influence of ultraviolet light. (See section 4.6.)

Naphthalene reacts with maleic anhydride at 100° and 9500 atm. pressure to give the adduct across the 1,4-positions in 78% yield. 1-Nitronaphthalene, under similar conditions, undergoes 5,8-addition[61], whereas 2,3-dimethylnaphthalene in dichloromethane in the presence or absence of aluminium chloride undergoes 1,4-addition[62]. 1,2,3,4-Tetramethylnaphthalene reacts with a thirty-fold excess of maleic anhydride in boiling benzene to give 90% of the addition compound, but in boiling xylene, equimolar amounts of the reactants gave only 4.6–6.4% of the adduct, in equilibrium with the components[62a].

No successful additions to phenanthrene appear to have been reported (although again the photocatalysed reaction is known[63]), but anthracene, naphthacene, pentacene, *etc.* react with dienophiles with increasing ease[64]. The proportion of adduct in equilibrium with the reactants is a function of any substituents present in the arene, the temperature, and the solvent. Table 3 shows[65] the amount of the adduct which is in equilibrium with an equimolar amount, or with a thirty-fold molar excess of maleic anhydride in boiling xylene.

TABLE 3

DISSOCIATION OF ANTHRACENE–MALEIC ANHYDRIDE ADDUCTS

*Arene**	% Adduct at equilibrium in boiling xylene	
	1 mol maleic anhydride	30 mol maleic anhydride
Anthracene	99	—
9-Methylanthracene	99	—
9,10-Dimethylanthracene	98	—
9-Phenylanthracene	75	97
9,10-Diphenylanthracene	16	78
1,2-Benzanthracene	84	99
1,2–5,6-Dibenzanthracene	30	91

* The numbering of the ring is as follows

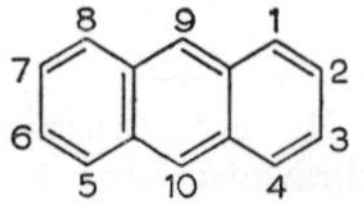

For preparative purposes it is often an advantage to use a large excess of one of the reagents, and to carry out the reaction at as low a temperature as possible. The reversibility of the reaction can often be put to advantage in preparing arenes and olefins, or for "protecting" a dienophile carbon–carbon double bond.

1. 6 Dimethylenecycloalkanes

Many dimethylenecycloalkanes will react readily with dienophiles, and the reaction can be used to identify the structure in unstable intermediates.

If allene and maleic anhydride are heated in benzene at 175°, the octahydronaphthalene derivative (XXX) is formed[66]. The intermediate 1,2-dimethylenecyclobutane (XXXI) has been isolated, and shown to give the mono-adduct (XXXII) with maleic anhydride[67] at 78°.

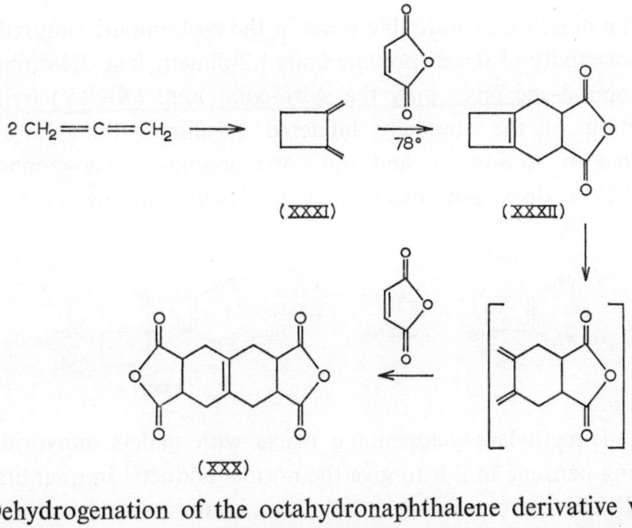

Dehydrogenation of the octahydronaphthalene derivative with bromine can be used as a route to functionally substituted naphthalenes. The hexacene (XXXIII) has similarly been prepared via the allene–naphtha-1,4-quinone adduct[68].

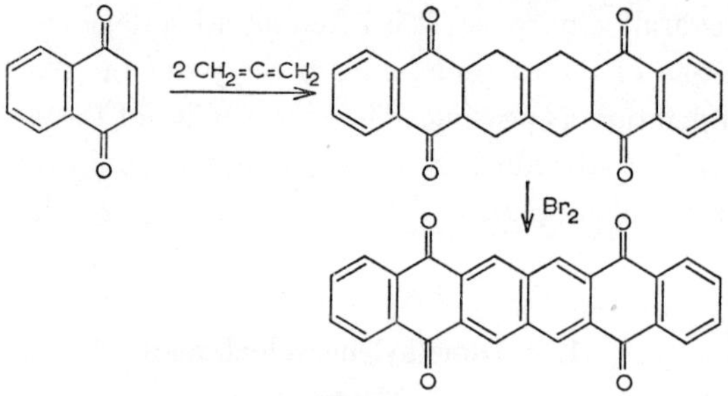

1,2-Dimethylene-3,4-diphenylcyclobutane reacts with one mole of maleic anhydride to give the adduct (XXXIV)[69].

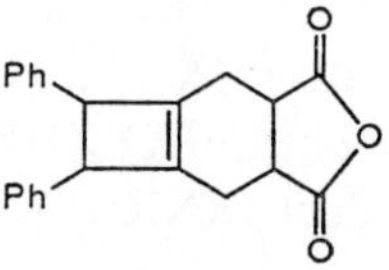

The presence of a double bond in the cyclobutane ring reduces the reactivity of the dienophile unit; 1,2-dimethylene-3,4-diphenyl-cyclobut-3-ene gives only the spiro-compound (XXXV) (*cf.*, the reaction of the sterically hindered 1,1-dimethylbuta-1,3-diene, quoted in section 2), and the corresponding benzo-compound (XXXVI) does not react with tetracyanoethylene in boiling toluene [70].

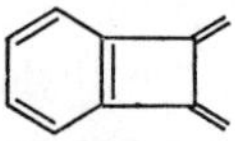

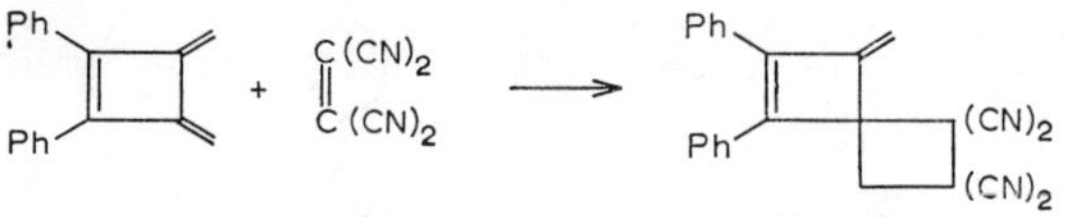

1,2-Dimethylenecyclopentane reacts with maleic anhydride in boiling benzene in 2 h to give the normal adduct[71] in quantitative yield. The corresponding cyclohexane is more reactive and adds exothermically to maleic anhydride in ether at room temperature[72], and the reaction with benzoquinone has been used as a route to pentacene (XXXVII)[73].

(XXXVII)

(16)

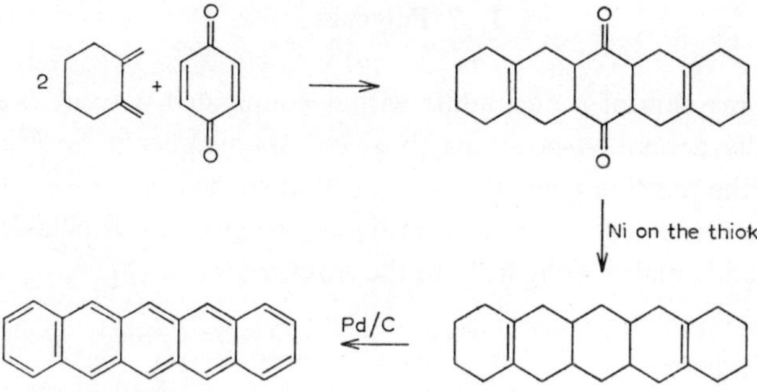

1,2,4,5-Tetramethylenecyclohexane is reported to react normally with two mol of dienophile[74], but hexaethylidenecyclohexane (XXXVIII), (hexamethylradialene), is inert to ethyl acetylene-dicarboxylate, maleic anhydride and quinone[75].

(XXXVIII)

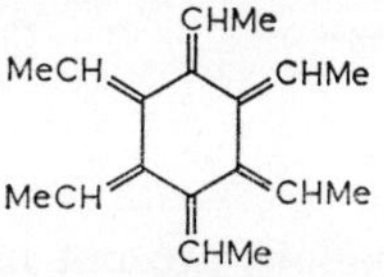

1,2-Dimethylenecyclohexa-3,5-diene (XL) can be obtained as an unstable intermediate from a number of sources, for example, by the pyrolysis of the sulphone (XXXIX). Its presence is inferred from its reaction with dienophiles (*e.g.* *N*-phenylmaleimide), which distinguishes it from its isomer, benzocyclobutene[76].

(XXXIX) (XL)

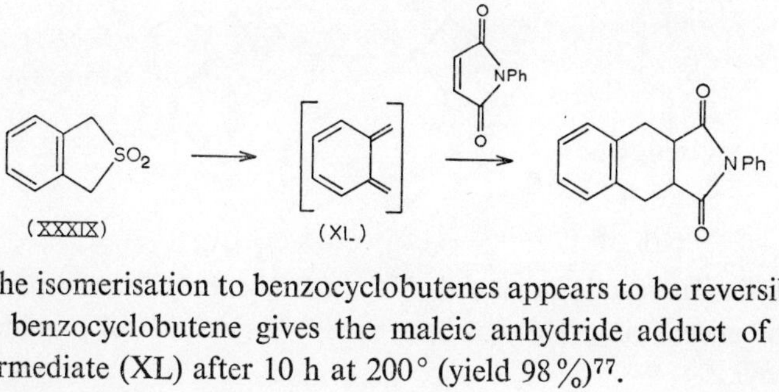

The isomerisation to benzocyclobutenes appears to be reversible, and benzocyclobutene gives the maleic anhydride adduct of the intermediate (XL) after 10 h at 200° (yield 98%)[77].

1. 7 Polyenes

The reaction of a dienophile with a conjugated polyene occurs normally across 1,4-positions to give a six-membered ring[78], the site of the reaction being determined by the usual principles which govern diene reactivity. For example, *trans-trans-cis*-octa-2,4,6-triene adds maleic anhydride to the *trans,trans* end (17)[79].

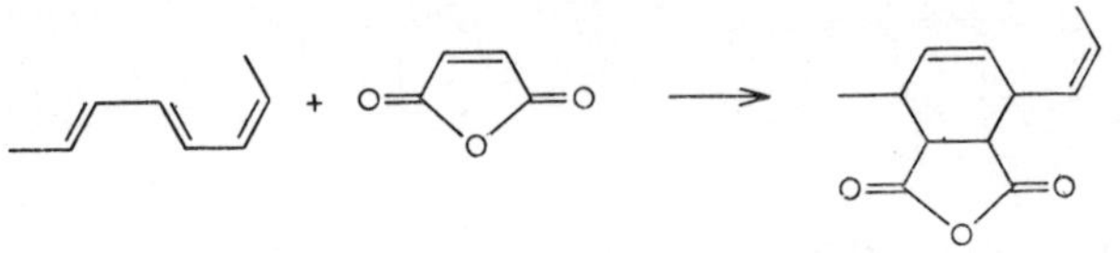

Again, in 1,1,8,8-tetraphenyloctatetraene the phenyl groups inhibit reaction of the terminal diene units, and addition to only the medial diene occurs[80].

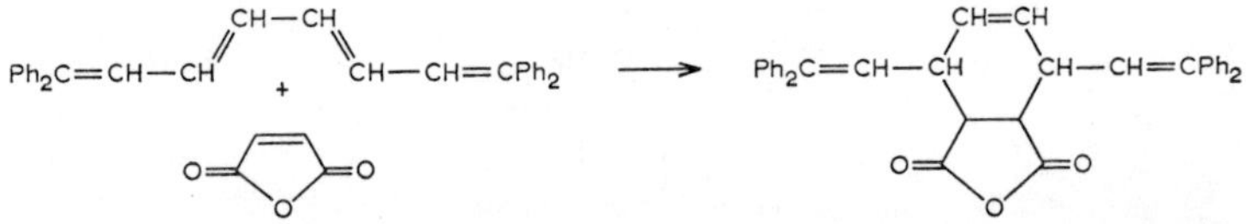

If the cisoid conformations are not hindered, a tetraene adds two molecules of maleic anhydride. For example, β-parinaric acid has been established, by its reaction with maleic anhydride, to be the all-trans compound (XLI)[81].

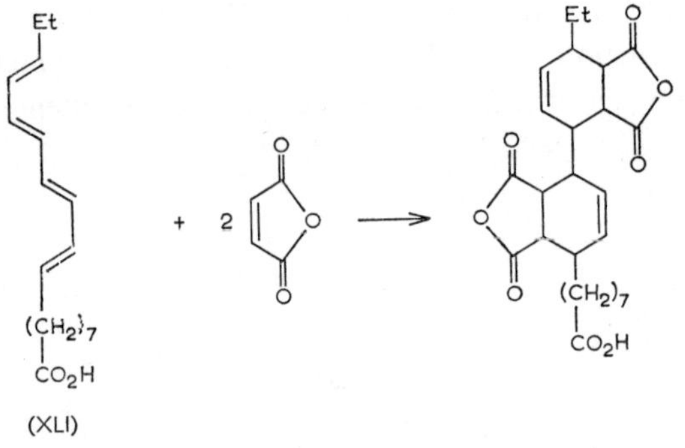

A similar reaction proceeds if the two diene units are separated by a benzene ring. This has been used[82] in synthesis of *p*-quinquephenyl.

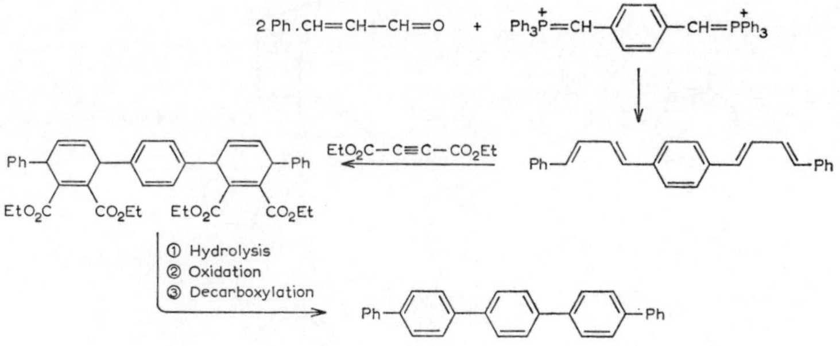

Cross-conjugated tetraenes again give the normal derivatives; thus 2,3-divinyl-1,3-butadiene reacts with, for example, 1,4-benzoquinone, as follows[83]:

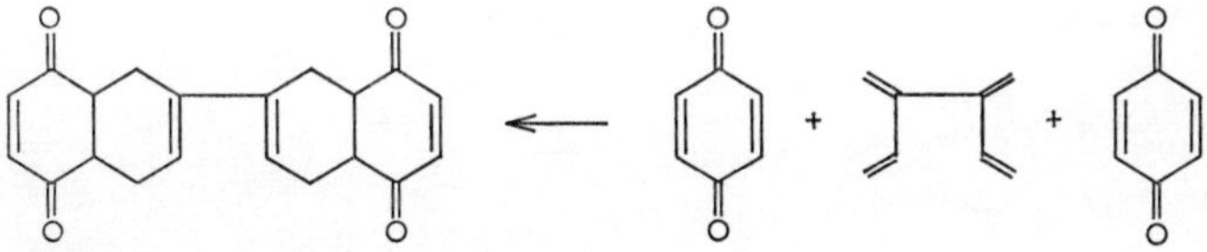

The Diels–Alder reaction can also be carried out if one of the conjugated double bonds is also part of a cumulene unit; the adduct then readily rearranges to give a methylarene, *e.g.*[84]:

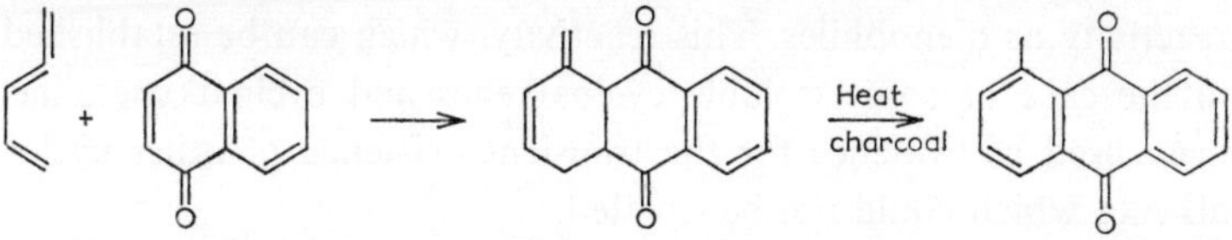

Tropone adds maleic anhydride across the 2,5-positions[85], but

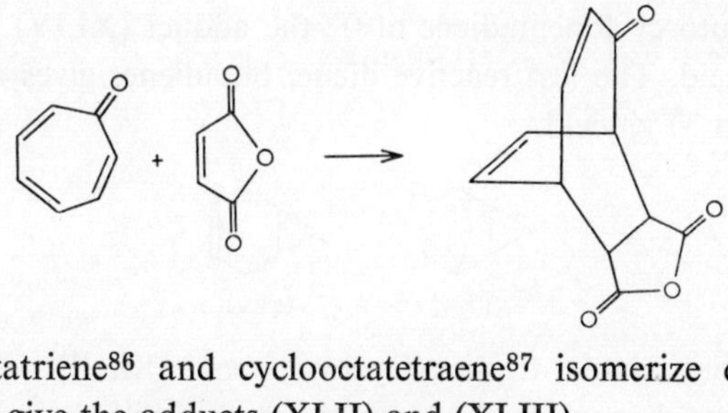

cycloheptatriene[86] and cyclooctatetraene[87] isomerize during reaction to give the adducts (XLII) and (XLIII).

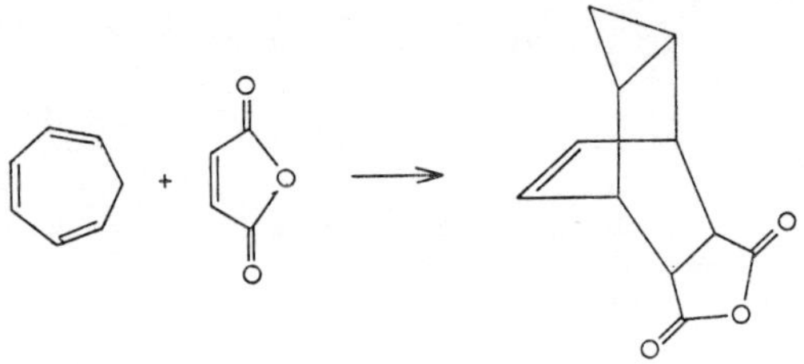

1.8 Cycloalkenes and cycloalkynes as dienophiles

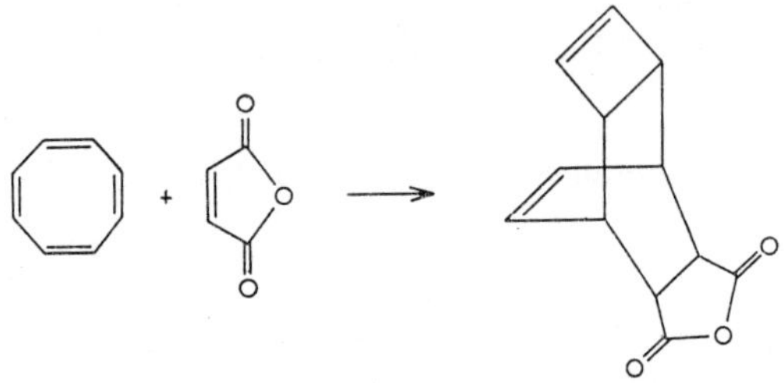

Steric strain in small-ring alkenes and alkynes reduces their thermodynamic stability, but at the same time enhances their reactivity as dienophiles. This reactivity, which can be established in the case of some isolable cycloalkenes and cycloalkynes, has been used as evidence for the transient existence of other cycloalkynes which could not be isolated.

Cyclopropene is stable as a solid at liquid nitrogen temperatures, but it slowly polymerises at —80°. If it is swept in a stream of nitrogen into cyclopentadiene at 0° the adduct (XLIV) is formed in 97% yield. The less reactive diene, butadiene, gives norcarane (XLV)[88] in 37% yield.

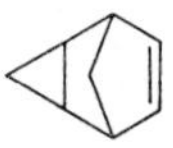

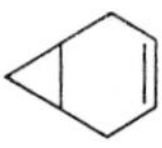

The more stable triphenylcyclopropene (XLVI) is a much less reactive dienophile than cyclopropene itself, but still much more reactive than *trans*-stilbene or tolane. It reacts with a twofold excess of tetracyclone in benzene to give the adduct (XLVII) in 88% yield after 6 days. This adduct, in acetonitrile solution at room temperature, slowly dissociates into the diene and dienophile; at higher temperatures the dissociation is in competition with decarbonylation to the cycloheptatriene (XLVIII)[89].

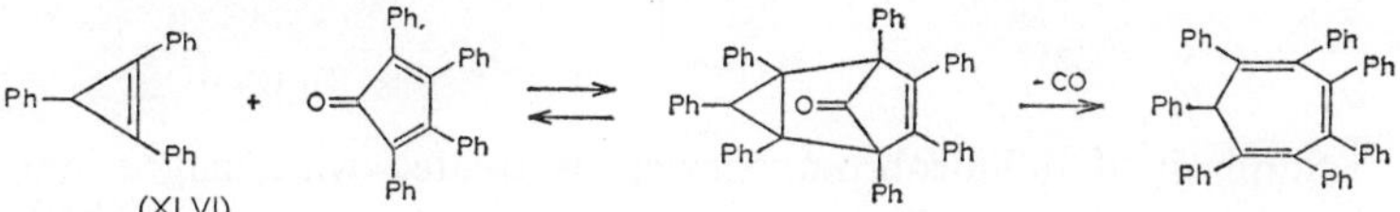

Cyclooctyne can be isolated as a stable liquid, but it reacts with 2,5-diphenyl-3,4-benzofuran to give the adduct (XLIX, $n = 6$) in 91% yield. Attempts to prepare cycloheptyne, cyclohexyne, and cyclopentyne by similar methods, as shown in (18), did not give isolable cycloalkynes, but their presence was inferred from the fact that, if the same reactive diene was present in the reaction mixture, the adducts (XLIX, $n = 3$–5) were obtained[90], in yields shown in Table 4 [91].

The transient existence of benzyne, "dehydrobenzene", a special and important example of a cycloalkyne, has been estab-

TABLE 4
YIELDS OF ADDUCTS (XLIX) IN %

Reactants	$n = 3$	$n = 4$	$n = 5$
Dibromide	2.1	50.5	64
Dihydrazone	0.5	7.0	26

lished by a number of different types of investigations[92–94]. Evidence that it could behave as a dienophile was first obtained as follows. *o*-Fluorobromobenzene in furan was treated with lithium amalgam giving the adduct (L) in 76% yield[95].

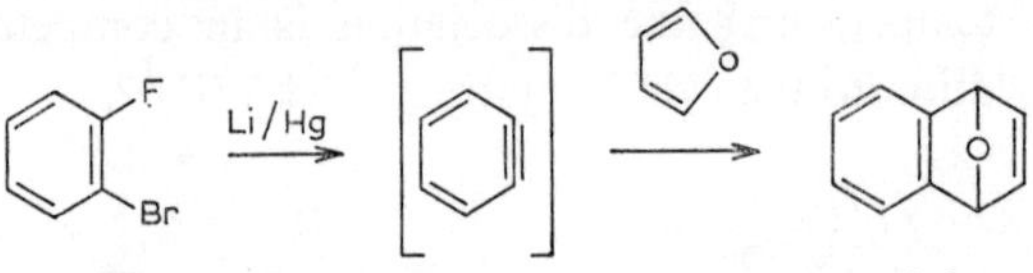

(L)

Similarly if *o*-fluorobromobenzene is treated with magnesium in the presence of anthracene, triptycene (LI) can be isolated in 28% yield[96], and a similar reaction in the presence of diphenylbenzofuran gives the adduct (LII)[97] (85% yield).

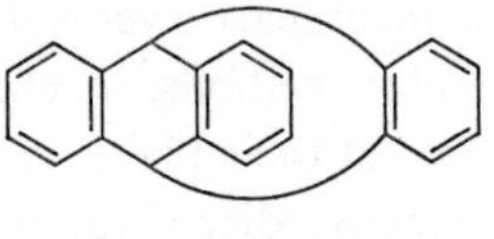

(LI)

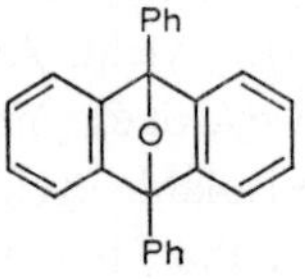

(LII)

Again the decomposition of the reagents (LIII-LV) in the presence of reactive dienes gives the adducts shown below[98,99].

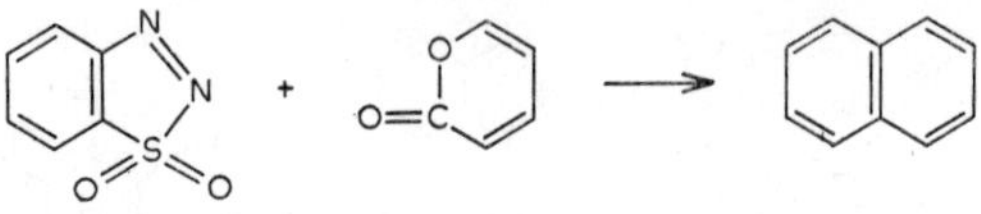

(LIII)

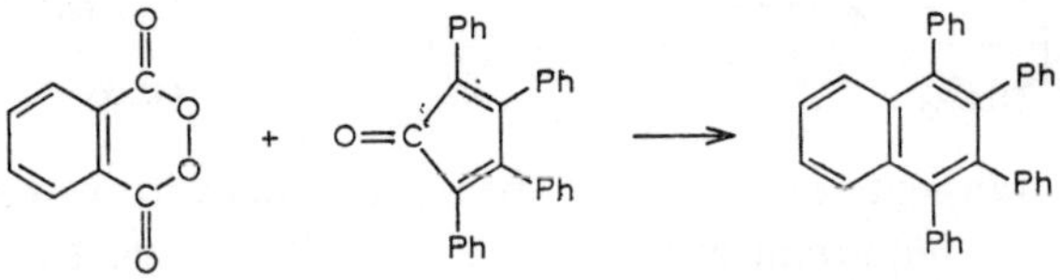

(LIV)

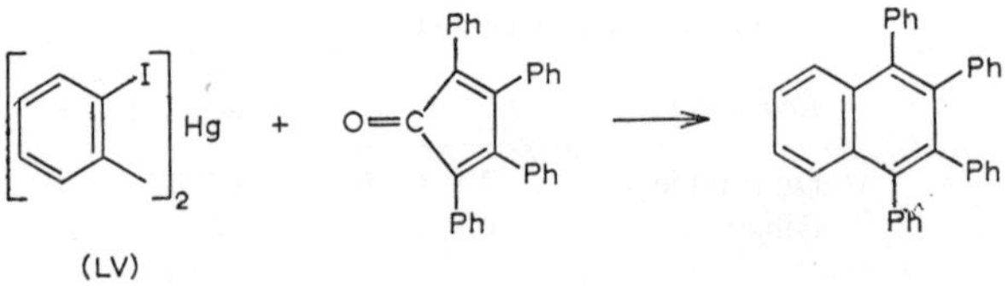

(LV)

It is reasonable to assume that benzyne is formed as a transient dienophile in all these reactions. Analogous reactions, shown below, can be brought about with derivatives of thiophene[100] and of pyridine[101], and it is tempting to suggest that the species thiophyne (LVI) and pyridyne (LVII) are now intermediates.

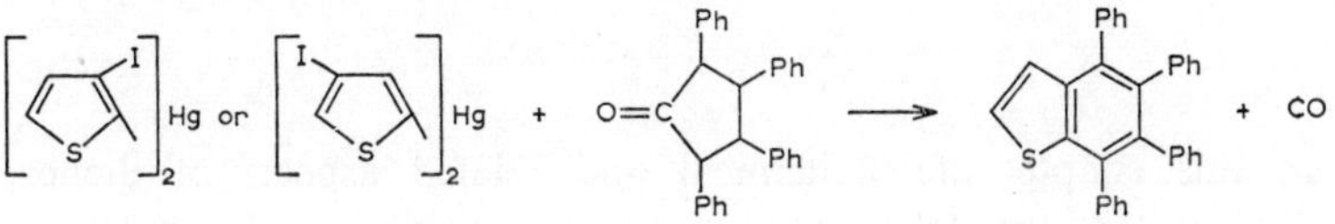

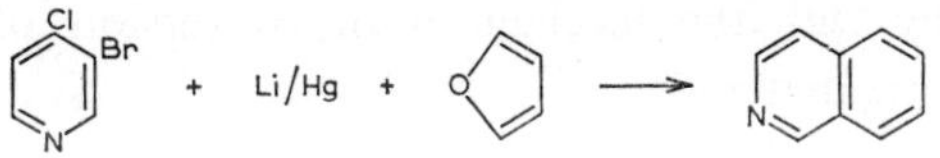

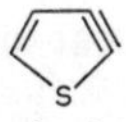

(LVI)

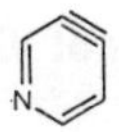

(LVII)

Chapter 2

STEREOCHEMISTRY AND RELATED TOPICS

In this chapter stereochemical and related aspects of dienes, dienophiles and Diels–Alder adducts are discussed, while the section 2,4 deals briefly with the mechanism of conversion of certain endo–exo-isomeric materials.

2. 1 Conformation of diene and retention of stereochemical integrity of dienophile

It has been pointed out in Chap. 1 that dienes can exist in cisoid and transoid conformations. It is known[102] that the constant, K, of equilibrium (3) depends on chemical conditions. Observations (see sections 1,2 and 1,3) show that cisoid dienes react relatively quickly with dienophiles, while transoid dienes are less reactive or give rise to the formation of polymers rather then Diels-Alder adducts. It seems clear, therefore (see some of the articles listed in ref. 1) that Diels–Alder adducts can only be obtained from *cis*-dienes or from dienes convertible into the cis conformation. Quantitative support for this important stereochemical principle is mentioned in section 5,1.

Another stereochemical principle relates to the dienophile. As an example we consider the reaction between cyclopentadiene and (a) dimethyl maleate or (b) dimethyl fumarate: these reactions can produce *cis*-adducts, (I) and (II), or a *trans*-adduct, (III). Experiments show that in (a) only *cis*-adducts and in (b) only the *trans*-adduct are formed, the two groups in the *cis*- or *trans*-position with regard to the ethylenic double bond of the dienophile remaining in the *cis*- or *trans*-position with regard to the ring in the associated species. This retention of stereochemical integrity of the dienophile has been observed in many other Diels–Alder reactions[1,43,103], provided the relative yield of the products is kinetically controlled. If this is not the case, the *cis*-adduct can be converted in a subsequent reaction into a thermodynamically more stable *trans*-isomer.

Rates of reaction between dienes and *cis-trans*-isomeric dienophiles have been investigated[104,105] and three situations have been observed: (a) the *cis*-dienophile reacts faster than the *trans*-isomer; (b) the *trans*-dienophile reacts faster than the *cis*-isomer; (c) the rates of addition of *cis*- and *trans*-dienophiles agree within the limits of the experimental errors. The reactants which were tested and the conditions of these experiments are indicated in Table 5.

TABLE 5

RELATIVE RATES OF DIELS–ALDER REACTIONS BETWEEN DIENES AND CIS–TRANS-ISOMERIC DIENOPHILES IN DIOXANE SOLUTION

Dienes	*Dienophiles*	*Temp.*	*Relative reactivity (see text)*	*Ref.*
Anthracene	Maleic acid, fumaric acid, citraconic acid, mesaconic acid	102°	(a)	104
Cyclopentadiene; 2,3-dimethyl-butadiene	*cis*- and *trans*-Benzoylbenzene-sulphonylethylene, *cis*- and *trans*-dibenzoylethylene, *cis*- and *trans*-bis-benzene-sulphonylethylene	40 and 100°	(b)	105
Cyclopentadiene; 9,10-dimethyl anthracene	Dimethyl maleate and fumarate	40 and 130°	(b)	105
2,3-Dimethyl-butadiene	Maleic and fumaric acid dinitrile	40 and 100°	(b)	105
Cyclopentadiene; 9,10-dimethyl anthracene	Maleic and fumaric acid dinitrile	40 and 130°	(c)	105

2. 2 Isomeric Diels–Alder adducts

The exo- and endo-isomeric Diels–Alder adducts produced from cyclic dienes and *cis*-dienophiles are of type (I) and (II) (*cf.* also section 1,1). With *trans*-dienophiles adducts of type (III)

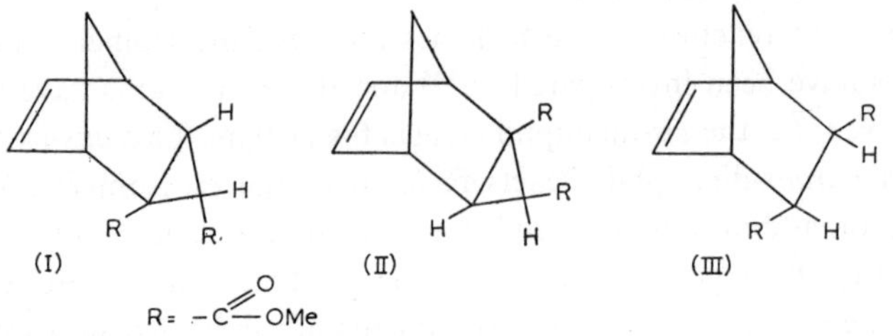

$$R = -\overset{O}{\underset{}{C}}-OMe$$

are obtained in which exo–endo isomerism is not possible. With open chain dienes

$$\begin{array}{c}R_1\\R_3\end{array}\!\!>\!\!C\!=\!CH\!-\!CH\!=\!C\!<\!\!\begin{array}{c}R_2\\R_4\end{array}$$

geometrically isomeric adducts have not been isolated, for the case in which $R_1 = R_3$ or $R_2 = R_4$, but dienes of the type

$$\begin{array}{c}R_1\\H\end{array}\!\!>\!\!C\!=\!CH\!-\!CH\!=\!C\!<\!\!\begin{array}{c}R_2\\H\end{array}$$

produce *cis–trans*-isomeric products, *e.g.*, (IV) and (V). Tables 6 and 7 show examples. Some of these associations are stereospecific, that is only the endo- or the *cis*-isomers are formed in detectable quantities, whilst in other cases both isomers are obtained in

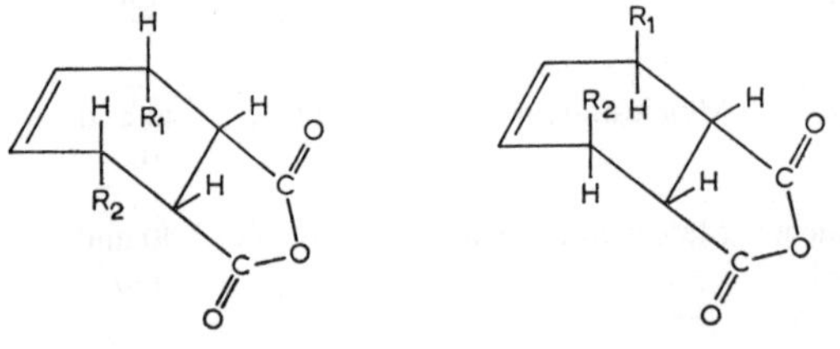

TABLE 6

FORMATION OF ENDO- AND EXO-ISOMERIC ADDUCTS IN DIELS–ALDER REACTIONS INVOLVING CYCLIC DIENES AND CYCLIC OR OPEN-CHAIN DIENOPHILES

Ref. no.	Diene	Dienophile	Solvent	Temp.	Reaction time (h)	Isomers formed*	Ref.
1	Cyclopentadiene	Maleic anhydride	C_6H_6	20–70°	½	only n	106
2	Cyclopentadiene	Cyclopentadiene	Pure liquid and gas phase	up to 100°	2–100	only n	107,108
3	Cyclopentadiene	Benzoquinone	Polar and non-polar solvents	20–50°	2.5–24	only n	41,107,109, 110
4	Cyclohexadiene		C_6H_6	20–50°	5	only n	106,107,111
5	Cycloheptadiene		Xylene	133°	5	only n	112
6	Cyclooctatetraene		Pure liquid	170°	½	only n	113
7	Dimethylfulvene	Maleic anhydride	Et_2O	0°	24	56%n; 44%x	114
			C_6H_6	80°	1	10%n; 90%x	
8	Diphenylfulvene		C_6H_6	20–80°	6–96	only n	115
			Xylene	133°	5	28%n; 12%x	
9	Furan		Et_2O	30°	5	only X	116
10	Furan	Maleic acid	H_2O	27.5°	23	only n	117
		$CH_2=CHR$					
11		R = Cl, Br and NH_2		170–180°	5–16	only n	11
12		R = COOH	Pure liquid	40°	?	mainly n	118
13		R = CN		100°	1	50%n; 50%x	119
14		R = COOMe		185–200°	10	81%n; 19%x	120
15	Cyclopentadiene	R = COOMe	12 Solvents, polar and non-polar	−35 to 170°		72–90%n; 2.8–10%x	121
16		Methyl methacrylate		0–139°	0.5–48	23–42%n; 77–58%x	
17		Methyl trans-crotonate		3–66°		46–70%n; 54–30%x	

* n = endo-adduct; x = exo-adduct; "only n" means that only the endo adduct could be isolated.

TABLE 7
FORMATION OF CIS- AND TRANS-ISOMERIC ADDUCTS IN DIELS–ALDERS REACTIONS INVOLVING OPEN-CHAIN DIENES $R_1HC=CH-CH=CH=CHR_2$ AND CYCLIC OR OPEN-CHAIN DIENOPHILES

Substituents in diene R_1	R_2	Dienophile	Solvent	Temp.	Reaction time (h)	Isomers formed	Ref.
COOMe	COOMe	Maleic anhydride	Pure liquid	150–160°	1	only *cis*	122
			C₆H₆	80°	20	only *cis*	123
			Xylene	133°	7	mainly *cis*	124
Me	H	Acrylic acid	Pure liquid	75–80°	2	only *cis*	} 125
Ph	Ph			130–135°	4	50% *cis*; 50% *trans*	
COOH	H						
Me	H	Acrylonitrile		130°	7	*cis* and *trans*	126
Ph	H			100°	12	mainly *trans*	127

yields which depend on experimental conditions. In connection with these observations the position of the equilibrium endo-isomer $\overset{K}{\rightleftarrows}$ exo-isomer is of interest. It has been suggested that certain exo-adducts are thermodynamically more stable than the endo-species (see, for instance, references 115, 117, 128) but reliable K values have not yet been determined. There are systems in which the ratio of the concentration of exo-adduct to the concentration of endo-adduct increases gradually with increasing time. It is not necessary, however, to conclude that this is due to a slow approach to equilibrium, because the effect may be brought about by irreversible consumption of the endo-isomer in other reactions. On the whole it appears reasonable to assume that neither the entropy nor the enthalpy changes are sufficiently large to shift these equilibria very far to either side. That the enthalpy differences between endo- and exo-adducts are indeed small has been confirmed by experiments[129]. In this situation it appears that the predominant formation of the endo-isomers, as in the typical cases listed in the first six lines of Table 6, is due to greater rate of formation, thermodynamic control of the product ratio being prevented because the relevant equilibria are "frozen in". If, on the other hand, both endo- and exo-isomers are produced, conclusions about relative rates of formation of these species are only permissible if careful tests, such as those described in ref. 121, have established that the relative yields of the products are not thermodynamically controlled. Similar considerations apply to the systems in Table 7; if both *cis*- and *trans*-isomers are obtained it must be proved, in the first instance, that the yields are not thermodynamically controlled before conclusions about different rates of formation can be drawn.

The stereospecific formation of endo-adducts is the experimental basis of the rule[43,130] of "maximum accumulation of unsaturation" which can be formulated as follows. The transition state of a Diels–Alder reaction is relatively stable if the mutual orientation of the addenda in the transition state is such that the distance between the "double bonds" is as small as possible, it being assumed that there is an electronic resemblance between initial and transition state*. Typical cases are the transition states shown in Fig. 3 and 4 in Chap. 5. It will be seen that the distances between the formal double bonds in structures A in these diagrams are relatively small if the orientation between the reactants is suitable for endo-addition. The rule was thought to be generally valid[43,130] but reactions have now been discovered in which it breaks down. Thus, in the associations 15–17 of Table 6 the rule requires predominant formation of the endo-adduct, while in fact both endo- and exo-isomers are formed, the product ratios being in all experiments kinetically controlled. This situation is further discussed in section 5,4.

Geometrically isomeric Diels–Alder adducts can also be obtained from maleic anhydride and naphthalene or anthracene derivatives. With disubstituted naphthalenes, for instance, species of type (VI) and (VII) have been prepared[131], in relative yields which depend on the nature of the substituents R and R'. More important, from a mechanistic point of view, are position isomeric adducts from acrolein or acrylates and mono-substituted butadienes. In

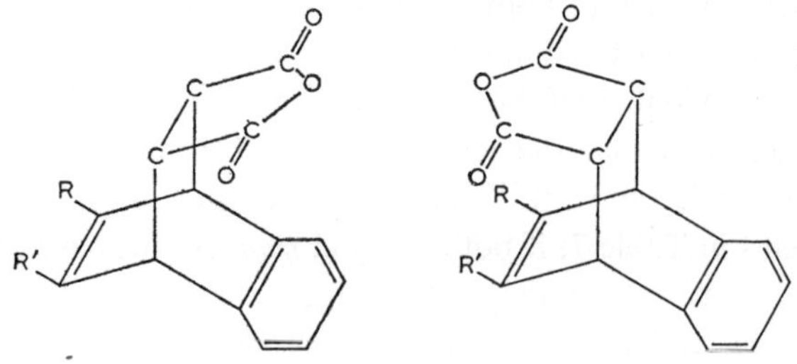

(VI)   (VII)

these associations 1- or 3-substituted dienes have been used; it has

---

* INGOLD has pointed out[4] (cf. also section 5,1) that electronic resemblance between initial and transition states of exothermic reactions is not incompatible with resemblance between the nuclear geometries of transition and final states.

been frequently observed that mainly *o*-products* are obtained in the former case and mainly *p*-products in the latter case.

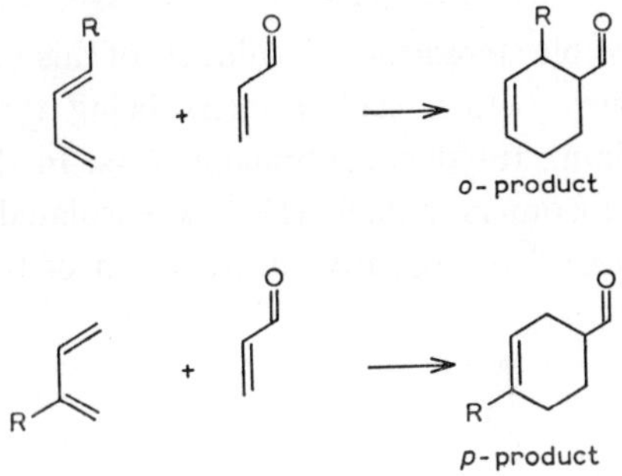

These observations were made for various substituents R, namely methyl[132], acetoxyl[133], alkyl[134], diethylamino[135], chloro[136], phenyl[127,137], cyano[138], carboxyl[125,139] and carboxymethyl[139a].

2. 3 Trimers and polymers

It is probable that the thermal polymerisation of cyclopentadiene is a step-wise Diels–Alder association, as pointed out in section 1,4. The trimer and similar adducts were investigated[140] by oxidative degradation and it was shown[140] that in every case the fused norcamphane ring is configurationally identical, but its actual stereochemistry could not be established. The problem became more interesting by the synthesis of insecticides, Aldrin and Dieldrin (see section 1,4), which also contain fused norcamphane rings[141]. The simplest adduct of this class is octahydrodimethanonaphthalene, (IX), which was prepared[142] from cyclopentadiene and the species (VIII), the latter being a by-product of the reaction between cyclopentadiene and ethylene[142].

---

* The prefixes "*o*" and "*p*" used here and in section 5,1 do not imply that these Diels–Alder adducts are benzenoid systems.

The four possible stereoisomeric adducts of this reaction (IXα)—(IXϑ) are shown below, such isomers being typical for other adducts containing fused norcamphane rings. In the case of (19) only one of the isomers, namely (IXβ) was isolated* the structure being established[142] by oxidative degradation of the product and

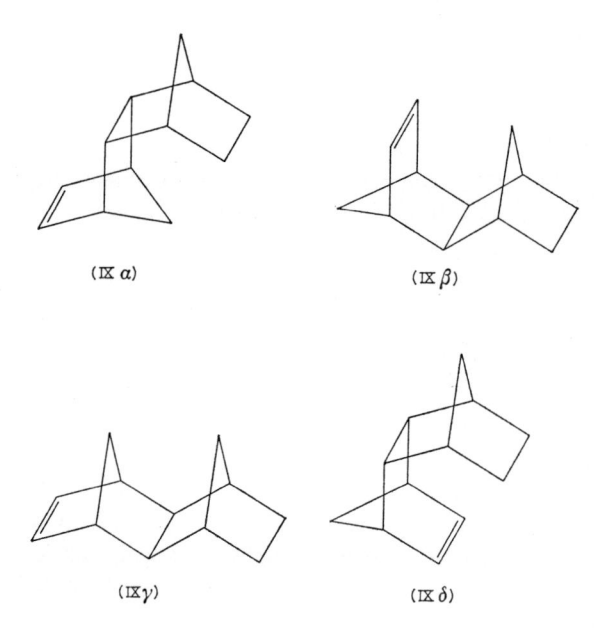

(IX α)

(IX β)

(IX γ)

(IX δ)

by other tests[142]. The stereochemical assignment is compatible with consideration of the four transition states shown in Fig. 9 (Chap. 5). It appears that this stereospecificity is essentially due to the same effects (to be discussed in Chap. 5) as those responsible for the predominant formation of endo-adducts in some of the asso-

---

* The value of the excellent work described in ref. 142 is somewhat decreased because solvent, temperature, reaction time and yields of adduct (IXβ) are not given.

ciations of Table 6. Steric hindrance is not of fundamental importance and it is misleading, therefore, to claim* that the course of association (19) and similar additions depends on "steric approach control".

Other adducts formed in two consecutive steps from butadiene or cyclopentadiene are below, the stepwise formation of species (XI) having been established by kinetic measurements[143].

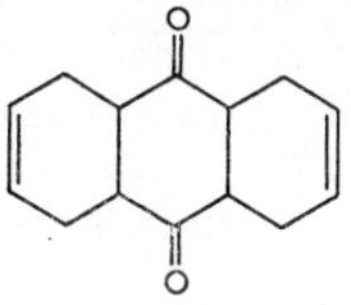

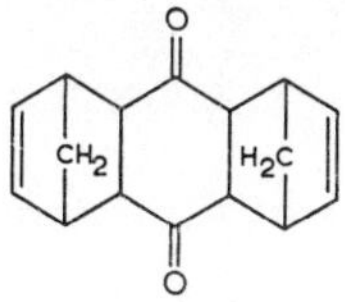

(X)   (XI)

These adducts can theoretically exist in the form of respectively four and six stereoisomers[144] but only the *cis*-anti-*cis*-compound (X') and the endo-*cis*-anti-endo-*cis*-isomer (XI'), were isolated[145,146].

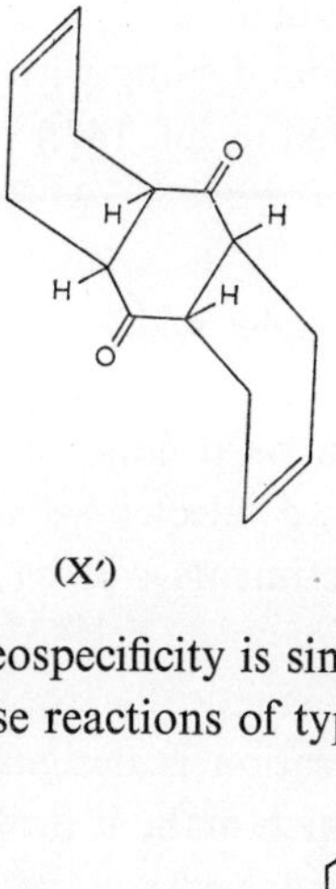

(X') 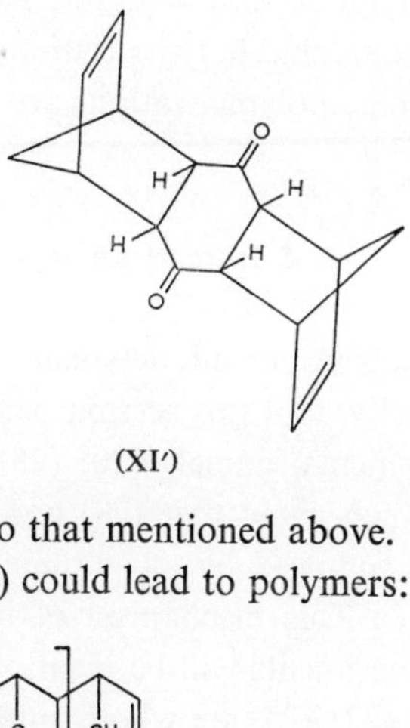   (XI')

This stereospecificity is similar to that mentioned above.

Stepwise reactions of type (19) could lead to polymers:

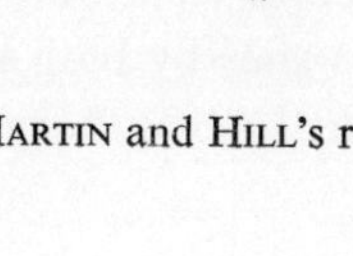

---

* See literature citations in MARTIN and HILL's review article[1].

If each step were characterised by the same stereospecificity as that in (19), there would be a spatial alternation of the norcamphane nuclei in the resulting polymer, such that corresponding planes in every other one would be parallel[142], and an overall

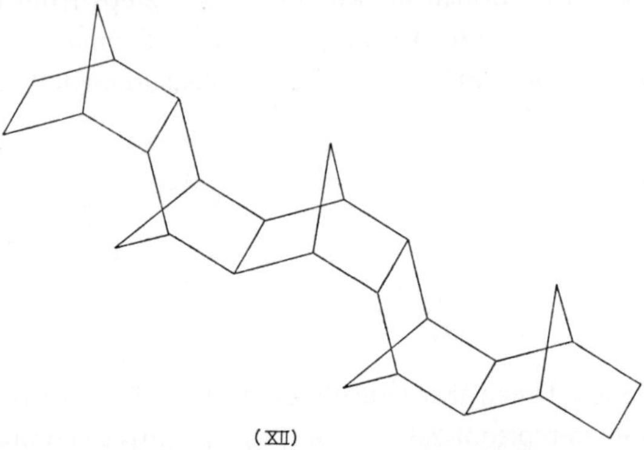

(XII)

linearity in space, as shown for the pentamer, (XII). If such reactions could be realized they would be examples of uncatalysed low molecular stereospecific polymerisations. (Papers dealing with catalysed stereospecific polymerisations are reviewed in ref. 147.)

2. 4 Isomerisation of Diels–Alder adducts

Many Diels–Alder adducts can be isomerised either catalytically or thermally, and this section deals with a selected set of thermal rearrangements, namely (20)–(25). Experiments will be described below which show that (20) and (21) involve Diels–Alder retrogressions followed by recombination of kinetically free diene and dienophile. This mechanism of isomerisation is designated as α. Other experiments will be mentioned which make it probable that reactions (22)–(24) are wholly intramolecular without participation of kinetically free dissociation products. This alternative mechanism of isomerisation is designated as β. Still other experimental observations indicate that (25) goes by both routes α and β simultaneously. It has been suggested[148] that the non-dissociative mechanism

β is of interest for an interpretation of the mechanism of thermal Diels–Alder retrogressions. This aspect will be briefly discussed in Chap. 5.

Reaction (20). Optically active adducts (XIII) and (XIV) were prepared[149], the required information to establish relative and absolute configuration being available[150]. On heating the optically active adducts (XIII), methyl-5-norbornene-2-exo-carboxylate and methyl-2-endo-methyl-5-norbornene-2-exo-carboxylate, in decalin, to 171°, completely racemized products (XIV), methyl-5-norbornene-2-endo-carboxylate and methyl-2-exo-methyl-5-norbornene-2-endo-carboxylate were obtained. Control tests[151] showed that this was not due to subsequent racemization of optically active species (XIV).

Reaction (21). The optically active enantiotropic adducts (XV) and (XVI) were prepared[152], as well as an optically active 9-phenylanthracene–maleic anhydride adduct, (XV′), which was ^{14}C-carbonyl labelled[152]. Two series of rate measurements were carried out[152] in *o*-dichlorobenzene at 132–174°; (a) the rate of racemizations of species (XV) and (XVI); (b) the Diels–Alder retrogression of (XV′) to produce 9-phenylanthracene and maleic

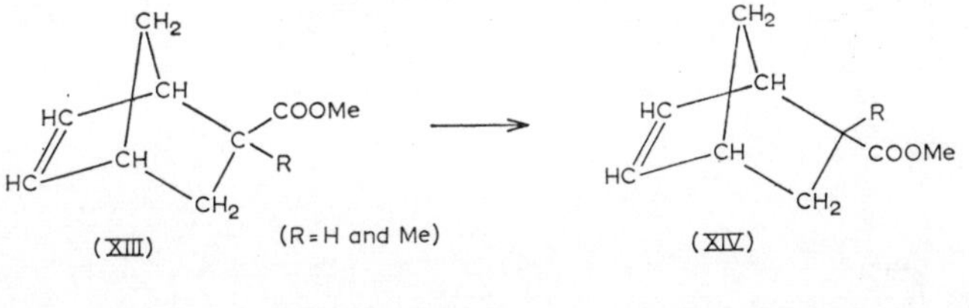

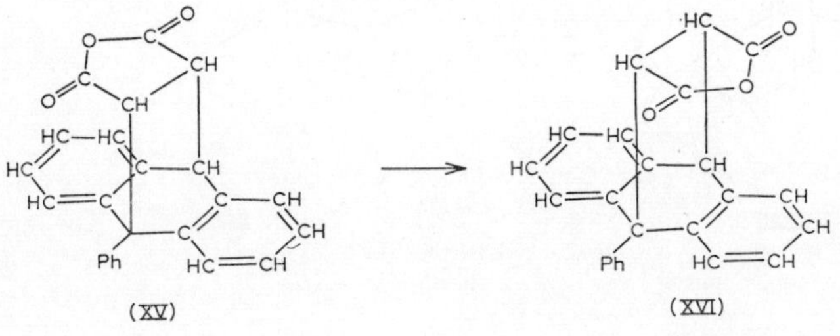

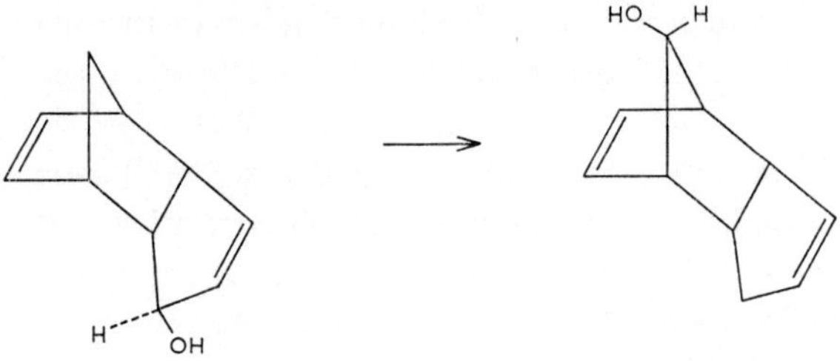

(XVII) → (XVIII)   (22)

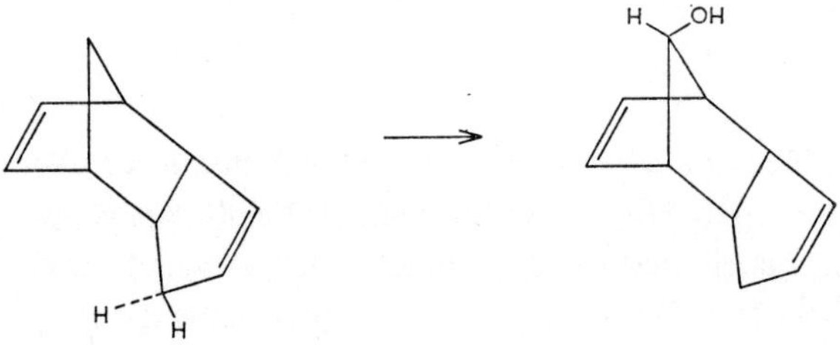

(XIX) → (XX)   (23)

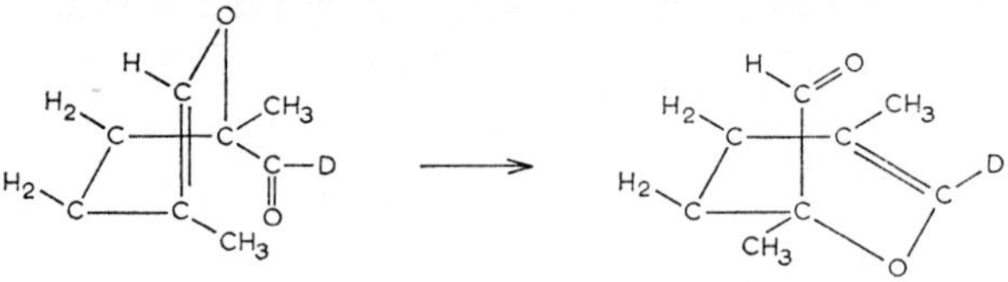

(XXI) → (XXII)   (24)

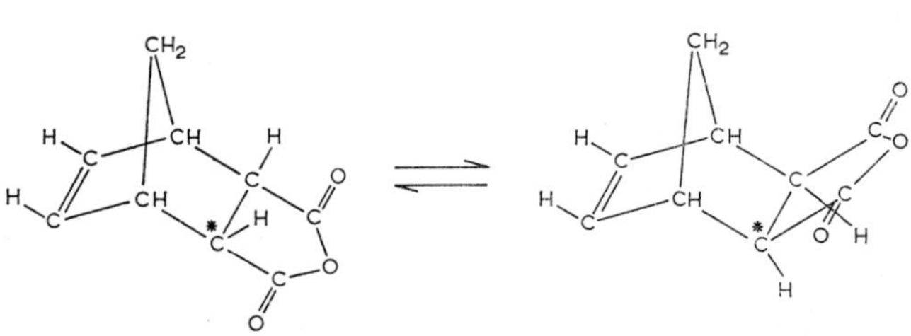

(XXIII) ⇌ (XXIV)   (25)

anhydride. It was found that the first-order rate coefficients obtained in (a) and (b) agreed within the limits of the experimental errors. These results are most simply explained[151,152] by the assumption that the dissociative mechanism α plays a role.

Reactions (22), (23), (24). The optically active species (XVII) and (XIX) were prepared[148] and heated to 135–150°; it was found[148] that these reactions are not only stereospecific in the sense that (XVII) produces only (XVIII), and (XIX) produces only (XX), but that there is also complete retention of optical purity*. Furthermore on heating the optically active deuterium labelled methacrolein dimer (XXI) to 171°, the species (XXII) is formed[153], there being again retention of optical purity.

These observations[148,153] are compatible with the intramolecular mechanism β and they cannot be explained by mechanism α, involving dissociated species which are stereochemically indeterminate. The rate of the reaction (23) in the pure liquid state has been investigated[153a] between 105 and 125° and it has been found that this rearrangement is unimolecular with activation parameters E and A of 32 ± 2 kcal/mole and about 10^{13} sec^{-1}. These values are similar to those in the third line of Table 19. Reaction (23) is retarded on increasing the hydrostatic pressure[153a], but significant conclusions about the activation volume are not possible.

The non-dissociative mechanism β can be compared with that playing a role in Cope rearrangements[154,155].

Thus it has been found[155] that the isomerisations of crotyl-isopropylmalonic ester and of crotyl-2-(2-butenyl)-cyanoacetic ester are intramolecular and occur with specific structural inversion of the crotyl group to α-methylallyl. The influence of hydrostatic

---

* For similar experiments with 1-ketodicyclopentadiene see ref. 148a.

pressure on the rate of Cope rearrangements involving cyclic molecules has not yet been investigated, but open chain systems[155a] are accelerated on increase of pressure.

Reaction (25). Adducts (XXIII) and (XXIV), labelled with ^{14}C in the positions indicated by asterisks[156] were dissolved in decalin, together with non-labelled maleic anhydride and heated to 188°. The distribution of ^{14}C of the products indicated[156] that both mechanism α and β play a role. In another investigation[157] the non-labelled adduct (XXIII) was heated in decalin (178–187°), the dissociated cyclopentadiene being "trapped". Attempts were made to estimate velocity coefficients k_α and k_β, relating respectively to the dissociative and intramolecular mechanisms. It turned out, however, that the numerical values of A_α and A_β of the rate equation were of a different order of magnitude than those characteristic of unimolecular processes (10^{11}–10^{14} sec^{-1}, see Table 19). It is not possible, therefore, to draw significant conclusions from the results of these tests, but further experiments of this kind, using an improved technique should be of interest. It would be worth while, for instance, to apply an experimental technique described in ref. 178, that is to trap the cyclopentadiene after removal from the reaction mixture.

Chapter 3

EQUILIBRIA

Diels–Alder reactions are reversible associations

$$a + b \underset{k_1}{\overset{k_2}{\rightleftharpoons}} c \tag{26}$$

where a and b are diene and dienophile, c is the adduct and k_2 and k_1 are velocity coefficients. These diene syntheses are frequently exothermic when the equilibrium, at room temperature, is far on the right hand side and is shifted towards the addenda on increase of temperature. In the gas phase, the equilibrium constant of (26) is defined by

$$K^g = \frac{k_2}{k_1} = \frac{[c]}{[a][b]} \tag{27}$$

where the expressions in square brackets are equilibrium concentrations. In a solvent, m, the equilibrium constant, K^m, is related to K^g by the Van 't Hoff–Dimroth relation[158]

$$K^m = K^g \frac{\sigma_c}{\sigma_a \sigma_b} \tag{28}$$

The solubilities, σ, are defined by

$$\sigma = \frac{\text{equilibrium concentration in solvent m}}{\text{equilibrium concentration in the gas phase}}$$

the subscript indicating here and below that the relevant quantities relate to a, b, or c of (26). In (28) the solubilities are the counterparts of activity coefficients in ionic equilibria. In solvents m and n the equilibrium constants K^m and K^n of a Diels–Alder reaction are related to each other by

$$K^m/K^n = \frac{\sigma_a^m \sigma_b^m \sigma_c^n}{\sigma_a^n \sigma_b^n \sigma_c^m}; \tag{29}$$

where σ^m and σ^n are the solubilities in solvents m and n.

The diene reactions here considered are unusual in that kinetics and equilibria can be studied in the gas phase and in polar or non-polar solvents. This field of research is not yet fully exploited and the few, mostly not very accurate equilibrium measurements now to be discussed should be regarded as the beginning of a more coherent story. One needs a systematic investigation into the influence of chemical conditions on enthalpy and entropy changes and a link up with activation energies and non-exponential factors of the Arrhenius rate equation. It would also be of interest to measure heat capacities of the participants of gaseous Diels–Alder reactions in order to gain a better understanding of entropy changes and of the contributions of vibrational partition functions.

3. 1 Theoretical relationships and results

In the case of gaseous associations the equilibrium constant, K^g, can be related to $\triangle E_0$, the work required to split the adduct c into the addenda a and b, all species being at rest and at infinite separation; K^g can also be related to an energy term $\triangle E_0'$ and to the heat and entropy changes $\triangle H$ and $\triangle S$; the relevant relationships are

$$K^g = W e^{-\triangle E_0 /RT} \tag{30}$$

$$K^g = W' e^{-\triangle E_0' /RT} \tag{31}$$

$$K^g = W'' e^{-\triangle H/RT} = e^{\triangle S/R} e^{-\triangle H/RT} \tag{32}$$

$$\triangle E_0 = \triangle H - \int_0^T \triangle c_p \, dT \tag{33}$$

$$\triangle E_0' = \triangle H + 4RT \tag{34}$$

It will be seen that $\triangle E_0$ would equal $\triangle E_0'$ if the vibrational heat capacities of initial and final state agreed with each other. The parameters W and W' in (30) and (31) can be represented by [158a]

$$W = \frac{N \lambda_c r_c}{1000 \lambda_a \lambda_b r_a r_b} \varrho = w\varrho \tag{35}$$

$$W' = w\varrho' \tag{36}$$

where λ is the translational partition function without volume factor, r is the rotational partition function and ϱ is the ratio of vibrational partition functions, q, defined by $\varrho = q_c/q_a \cdot q_b$, each partition function λ, r and q referring to an energy zero, which is that of the lowest state of the relevant degree of freedom*. One has furthermore

$$\log \varrho' - \log \varrho = \log e \left\{ \Delta \Sigma \left(\frac{h\nu/kT}{\exp(h\nu/kT) - 1} \right) \right\} \quad (37)$$

where the ν are the frequencies of the fundamental vibrational modes of the participants of (26). If some of the ν values of the adduct c are much lower than those of the addenda, a and b, ϱ and $\log \varrho' - \log \varrho$ may be $\gg 1$.

For comparison of gaseous and liquid phase equilibria, the solubilities σ in (29) are related to entropies and heats of solutions, ΔS^s and ΔH^s by

$$\sigma_a = \exp(\Delta S_a^s/R) \exp(-\Delta H_a^s/RT) = W_a^s \exp(-\Delta H_a^s/RT) \quad (38)$$

the expressions for σ_b and σ_c being analogous. Introducing (38) and its analogues into (28) and (29) and taking (32) into account one obtains

$$\Delta H^m - \Delta H^g = \Delta H_c^s - \Delta H_a^s - \Delta H_b^s \quad (39)$$

$$W''^{(m)}/W''^{(g)} = W_c^s/W_a^s \cdot W_b^s \quad (40)$$

$$\Delta H^m - \Delta H^n =$$
$$\Delta H_c^{s(n)} + \Delta H_a^{s(m)} + \Delta H_b^{s(m)} - \Delta H_c^{s(m)} -$$
$$\Delta H_a^{s(n)} - \Delta H_b^{s(n)} \quad (41)$$

$$\log W''^{(m)} - \log W''^{(n)} =$$
$$\log W_c^{s(n)} + \log W_a^{s(m)} + \log W_b^{s(m)} - \log W_c^{s(m)}$$
$$- \log W_a^{s(n)} - \log W_b^{s(n)} \quad (42)$$

The former two relations will be used for a comparison of gaseous and liquid phase equilibria and the latter two are of interest for effects discussed in the following chapter. Results of equilibrium and solubility measurements are in Tables 8 and 9.

---

* Standard formulae for calculating the contributions of translational, rotational, and vibrational degrees of freedom are given in the Appendix.

TABLE 8

HEAT AND ENTROPY CHANGES OF DIELS–ALDER REACTIONS IN THE GASEOUS AND CONDENSED STATE

(In calculating $\Delta S/R$ the concentration unit of the standard state was mole/l).

Ref. No.	*Addenda*	*Formula of adduct*	*Measurements in*	*Temp. range*	*—ΔH (kcal/mole)*	*—ΔS/R*	*Experimental technique*
1	Butadiene; ethylene	(I)	Gas	585–648°	30.0 ± 3.0	12.4 ± 4.1	a
2	2 Cyclopentadiene	(II)	Gas	—1 to 172°	17.0 ± 1.1	15.9 ± 1.6	b
3	2 Cyclopentadiene	(II)	Paraffin	—1 to 172°	16.3 ± 0.7	13.6 ± 0.9	a
4	2 Cyclopentadiene	(II)	Pure liquid	—2 to 80°	19.1 ± 1.7	18.0 ± 2.1	a
5	Cyclopentadiene; acrolein	(III)	Gas	108–242°	19.3 ± 1.9	14.5 ± 3.0	a
6	Cyclopentadiene; benzoquinone	(IV)	C_6H_6	8–79°	17.4 ± 1.9	14.5 ± 3.0	a
7	Cyclopentadiene; napthaquinone	(V)	C_6H_6	11–79°	16.0 ± 2.2	14.3 ± 4.4	a
8	9–Phenylanthracene; maleic anhydride	(VI)	*o*-Dichlorobenzene	152–175°	20.1 ± 1.5	19.1 ± 1.4	a,c

Experimental techniques: a) Kinetic measurements, see Tables 13, 14 and 19 and ref. 143, 152, 159; b) Kinetic and solubility measurements using (39) and (40); c) Estimation of steady state concentrations; see ref. 152.

(I)

(II)

(III)

(IV)

(V)

(VI)

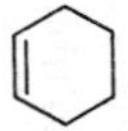

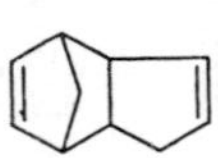

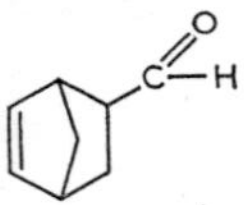

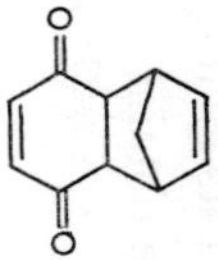

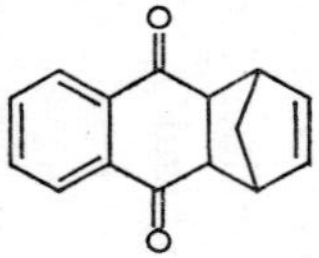

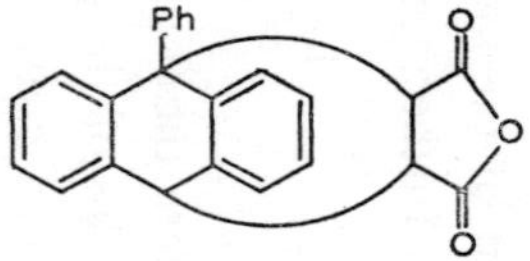

TABLE 9
HEATS OF SOLUTION, ΔH^s, AND W^s FACTORS OF CYCLOPENTADIENE AND ENDO-DICYCLOPENTADIENE RELATING TO THE EQUILIBRIUM BETWEEN THE GAS PHASE AND PARAFFIN SOLUTION (see (38) in text, and ref. 159)

	ΔH^s(kcal/mole)	$-\log W^s$	Temp. range
Cyclopentadiene	-7.5 ± 0.2	3.1 ± 0.1	27–73°
Endo-Dicyclopentadiene	-13.5 ± 0.3	4.9 ± 0.2	55–111°

3.2 Heats of reaction

The $\triangle H$ value of reaction ref. no. 1 in Table 8, is 13 ± 3 or 11 ± 4 kcal larger than the heats of reaction ref. nos 2 or 5. In order to explain this the energy term, $\triangle E_0$, is expressed by

$$-\triangle E_0 = 2\alpha - [2(\beta - \alpha) + \triangle \gamma - \triangle \delta] \quad (43)$$

where α and β are energies of formation of carbon–carbon single and double bonds and γ and δ are resonance energies and ring strains, $\triangle \gamma$ and $\triangle \delta$ being $\gamma_a + \gamma_b - \gamma_c$ and $\delta_a + \delta_b - \delta_c$. If it is assumed, as a first approximation, that $\triangle E_0 = E'_0$ and that ring strain, if any, in cyclohexene and cyclopentadiene is negligible it follows from (43) that

$$-\triangle H = 2\alpha - [2(\beta - \alpha) + \gamma_a] - 4RT \quad (44)$$
$$-\triangle H = 2\alpha - [2(\beta - \alpha) + \gamma_a + \gamma_b + \delta_c] - 4RT \quad (45)$$

the former equation relating to reaction ref. no. 1 in Table 8 and the latter to reactions ref. nos. 2 and 5. The resonance energies of butadiene, cyclopentadiene and acrolein can be taken[160] to be 3–6 kcal/mole and the ring strain[129] in the adducts (II) and (III), produced by the endo-methylene bridge 6–7 kcal/mole. It follows, therefore from (44) and (45) that the heat of reaction, ref. no. 1 in Table 8, should be 9–13 kcal larger than the $\triangle H$ values of ref. nos. 2 and 5. This is compatible with the experimentally observed difference. All the reactions listed in Table 8 are exothermic, notwithstanding the decrease of resonance energy, in passing from the initial to the final state. In Diels–Alder additions to certain aromatic hydrocarbons, the situation can be quite different, as shown by an empirical[160] and molecular orbital method[161]. In the hypothetical reactions (46)–(48), for instance, $\Delta \gamma$ is 24–37 kcal/mole (see Table 7.2 of ref. 160) which is sufficiently large to make it improbable that the adducts are thermodynamically stable. In (49), on the other hand, in which the adduct is actually formed[20], the calculated[161] $\Delta \gamma$ values are much smaller, namely 12–16 kcal/mole.

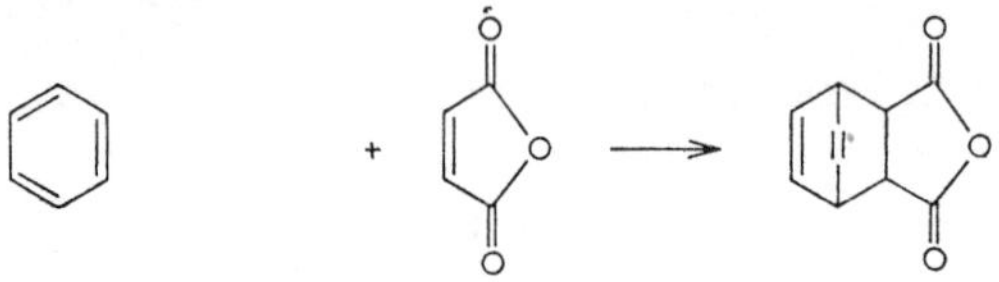

(46)

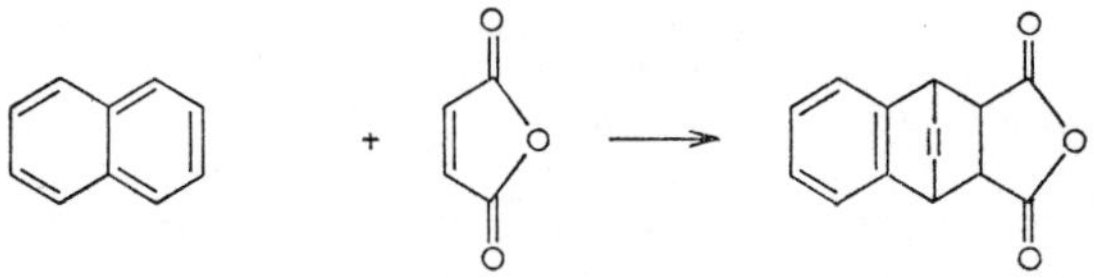

(47)

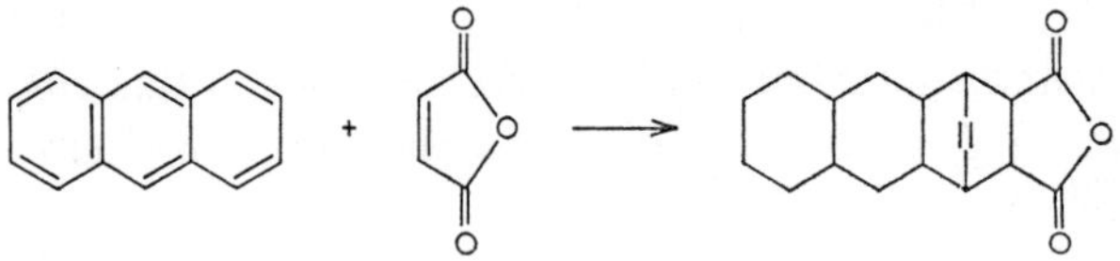

(48)

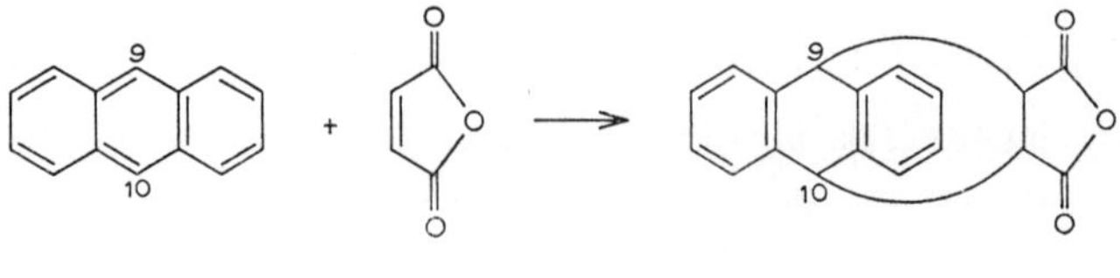

(49)

3. 3 Entropies of reaction

The entropy change, ΔS, of typical Diels–Alder reactions can be derived from the figures in the seventh column of Table 8. All ΔS values are negative as one would expect for such associations. The entropy changes of the gaseous reactions ref. nos 1,2 and 5 are less negative, however, than for associations of rigid species of comparable size and symmetry, where the entropy change would be entirely due to loss of translational and rotational degrees of freedom. The enhanced statistical probability of the products (I), (II) and (III) is due to the influence of vibrational partition functions[162,163]. Numerical values of individual vibrational partition functions are not known, but the ratio ϱ' of (36) has been computed[162,163] and numerical values are listed in Table 10. The value of this ratio is at least 200, which indicates[162] that the frequencies, ν, of some vibrational modes of the adducts are lower than the lowest ν values of the addenda. Such modes of particularly low frequency may be due to the bending of bonds joining certain ring carbon atoms together or to the twisting of ring carbons or hydrogen atoms relative to the rest of the molecule. In the adducts (II) and (III) vibrational modes have also to be considered[162] in which polyatomic parts of these molecules move relative to each other. Such modes might be inactive in the Raman or infrared spectrum and their frequencies could be lower than the lowest spectroscopically observed values.

TABLE 10

VIBRATIONAL CONTRIBUTIONS, ϱ', DEFINED BY (36), OF THREE GASEOUS DIELS–ALDER REACTIONS

Ref. no. of association in Table 8	1	2	5
log ϱ'	5.3 ± 3.0	3.3 ± 0.9	5.2 ± 1.7

3. 4 Equilibria in solution

It follows from the figures in the second column of Table 9 in conjunction with (39) that the association of cyclopentadiene is more exothermic in the gas phase than in paraffin solution. This is because the process of dissolving two moles of cyclopentadiene is more exothermic than the dissolution of one mole dicyclopentadiene. The difference $\triangle H^g - \triangle H^m$ (m = paraffin) is only 1.5 ± 0.7 kcal.mole, which is small compared to either $\triangle H^g$ or $\triangle H^m$.

Introducing the figures in the third column of Table 9 in (40) one obtains a $\log_{10}(W''^{(m)}/W''^{(g)})$ value of 1.3 ± 0.4. This relatively small ratio is due to the fact that the statistical weight factor W^s of dicyclopentadiene is considerably smaller than the factor W^s of cyclopentadiene, implying that dicyclopentadiene is characterized by a more unfavourable entropy of solution than cyclopentadiene. If the W^s factors of both species were identical, e.g., 10^{-3}, the ratio $W''^{(m)}/W''^{(g)}$ would be about 10^3 times larger than the experimentally observed ratio. The relatively close agreement between $\triangle H_c^s$ and $2 \triangle H_a^s$ and between W_c^s and $(W_a^s)^2$ may be due to the fact that cyclopentadiene, dicyclopentadiene and solvent paraffin are hydrocarbons. It is unlikely that in the case of polar reactants and solvents the effects are as small as those indicated by the $\triangle H$ and $\triangle S$ values in ref. nos. 2–5 of Table 8.

3. 5 Mode of alignment of diene and dienophile

In the majority of reactions between a diene and a dienophile Diels–Alder adducts are formed according to (1), where the cyclic species may contain an endomethylene bridge. An alternative mode of alignment leads to cyclobutane derivatives, as shown below.

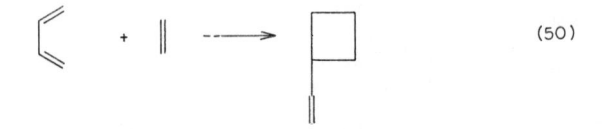

(50)

A few typical reactions of this type (50) are listed in Table 11. The $\triangle E_0$ values of reactions ref. nos. 2 and 5 in Table 8 and of associations of type (50) should be similar because the ring strain in the bridged six-membered rings of adducts (II) and (III) and of cyclobutane is similar[168]. There is no reason to suppose that, in general, the entropy changes of (1) and (50) are markedly different and it can be concluded, therefore, that the frequently observed Diels–Alder mode of alignment is not due to thermo-

TABLE 11
TYPICAL REACTIONS IN WHICH DIENES AND DIENOPHILES PRODUCE CYCLOBUTANE DERIVATIVES

Diene	Dienophile	Reaction conditions	Ref.
Butadiene	Butadiene	150°, 100 atm, 18 h	164
Butadiene	Tetrafluoro-	125°, 8 h	165
Cyclopentadiene	ethylene	470–480°, 1 atm, heterogeneous catalyst	166
4-Methyl-1,3-pentadiene	Tetracyano-ethylene	2–57°, tetrahydrofuran solution	167

dynamic instability of a 1,2-addition product*. The occurrence of associations of type (1) rather than (50), under mild reaction conditions, must be due to a relatively low activation energy of the former reaction, owing to effects to be discussed in Chap. 5.

---

* For mechanistic aspects of cycloadditions of type (50) see ref. 168a.

Chapter 4

KINETICS AND CATALYSIS

A prerequisite for an understanding of the mechanism of Diels–Alder associations and retrogressions is a knowledge of the kinetics. For this reason pertinent results relating to thermal non-catalysed diene reactions are given in sections 4,1–4,4. while work on catalysed and photochemical Diels–Alder addition is reported in sections 4,5 and 4,6. A discussion of some of these investigations is reserved for the following chapter.

A number of Tables contain non-exponential A factors and activation energies, E, of the Arrhenius rate equation. The A factors of bimolecular associations, with unit concentration as the standard state, are related to entropies of activation, $\triangle S^*$, by[168b]:

$$A = \varkappa \, (kT/h) e^{x} e^{\triangle S^*/R} \qquad (51)$$

where $\varkappa$ is the transmission coefficient and x is 2 or 1, according to whether the association occurs in the gas phase or in the condensed state. The relationship between A and $\triangle S^*$ is similar to that between the factor W'' and the entropy change $\triangle S$ in (32), both $\triangle S^*$ and $\triangle S$ depending on ratios of partition functions[168b]. In unimolecular gaseous or liquid phase Diels–Alder retrogressions, on the other hand, the exponent x in (51) is always 1 and moreover entropies of activation are independent of the concentration unit.

4. 1 Associations in the gaseous and condensed state

The unsaturated hydrocarbons listed in Table 12 react in the gas phase to produce mixtures of compounds the constitution of

TABLE 12

THERMAL GASEOUS ASSOCIATION REACTIONS;
REACTANTS AND EXPERIMENTAL CONDITIONS

Reactants	Temp. range	Range of initial pressure (mm Hg)	Ref.
Acetylene	400–650°	24–2,300	169,170
Ethylene	313–620°	760–200,000	170,171,172
Propylene	330–400°	50,000–180,000	172
Isobutylene	330–370°	70,000–200,000	172
2-Butylene	330–380°	70,000–220,000	172
Amylene	350–420°	120,000–150,000	172
Isoprene	255–371°	212–739	173
1,3-Pentadiene	279–419°	not given	174
2,3-Dimethyl-1,3-butadiene	309–400°		

which may depend on temperature and pressure. Some of these reactions are heterogeneous and influenced by traces of catalysts and it is not possible, therefore, to estimate kinetic parameters relating to bimolecular steps contributing to the overall processes. This unsatisfactory state of affairs is not characteristic of all gaseous associations involving polyatomic molecules. The dienes and dienophiles listed in Table 13 produce Diels–Alder adducts in the homogeneous gas phase and the initial stage of these reactions is kinetically of the second order. At higher conversions consecutive, simultaneous and reverse processes play a role, thereby complicating the kinetics. The figures in Table 13 relate to a stage where these complications are not significant.

A gas phase reaction leading to an adduct similar to that found in association (ref. no. 1) of Table 13 is as follows

$CH_2=CH-CH=CH-CH=CH_2 \longrightarrow$ [cyclohexadiene structure]

The kinetics of this cyclisation have been studied[177a] between 117 and 190°. The reaction is unimolecular with activation parameters $E = 29.9 \pm 0.5$ kcal: mole and $A \sim 10^{11}$ sec^{-1}.

TABLE 13
KINETIC PARAMETERS A AND E OF GASEOUS DIELS–ALDER REACTIONS

Ref. No.	Reactants	Temp. range	Range of initial pressure (mmHg)	$\log A$ ($A =$ l/mole sec)	E (kcal/mole)	Ref.
1	Butadiene; ethylene	487–698°	69– 691	7.5 ± 0.8	27.5 ± 1	163
2	2 Butadiene	173–386°	443–5000	6.96 ± 0.18	23.7 ± 0.2	175
3	2 Cyclopentadiene	79–150°	165– 732	6.1 ± 0.4	16.7 ± 0.6	176
4	Cyclopentadiene; acrolein	107–209°	42– 328	6.2 ± 0.5	15.2 ± 0.8	
5	Butadiene; acrolein	155–332°	43– 470	6.2 ± 0.4	19.7 ± 0.6	177
6	Isoprene; crotonaldehyde	242–300°	91– 440	6.0 ± 0.7	22.0 ± 1.0	
7	Isoprene; acrolein	218–333°	39– 432	6.0 ± 0.5	18.7 ± 0.8	

The reactions listed in Table 14 and other Diels–Alder reactions which have been investigated in solution are second order processes up to high conversion. The measurements were carried out at lower temperatures than in the gas phase and for this reason kinetic complications were not significant. Second order kinetics were not observed in the addition of dialkyl azodicarboxylates to cyclopentadiene[190] in methanol, ethanol, chloroform and acetic acid solution, possibly owing to an irreversible reaction between the solvents and this particular dienophile. A less serious complication has been observed[185] in the reaction between maleic anhydride or chloromaleic anhydride and anthracene or dimethylanthracene. It appears that there is a pre-equilibrium diene + dienophile $\rightleftharpoons$ coloured complex, but it is probable[185,152] that the coloured species is formed in a side reaction, without mechanistic significance.

The activation energies, E, listed in Table 13 and 14 are between 8 and 27 kcal and most of the log A values are between 4.0 and 7.5. Using (51), it follows that all the entropies of activation, $\triangle S^*$, are negative, as one would expect for associations in which translational and rotational degrees of freedom of reactants are converted into vibrational degrees of freedom of the transition states. Assuming that the transmission coefficient, $\varkappa$, in (51) is unity, numerical values of $\triangle S^*/R$ can be calculated: in the typical case of reaction, ref. no. 3, Table 13, $\triangle S^*/R$ is —17.6 $\pm$ 1.1, the concentration unit of the standard state being mole/l. This entropy change and that listed in the second line of Table 8, relating to the fully formed adduct, agree within the limits of the experimental errors. A similar agreement is characteristic of other Diels–Alder reactions (See Tables 8, 13 and 14).

From the point of view of the collision theory, the associations listed in Tables 13 and 14 are "slow reactions", both in the gaseous and condensed state, because the kinetic A factors are considerably smaller than the bimolecular collision frequency, $\sim 10^{11}$ l/ mole sec. From the point of view of the transition state theory, on the other hand, some of these reactions could be described as "fast" for reasons discussed in section 5,2.

TABLE 14
KINETIC PARAMETERS A AND E OF DIELS–ALDER REACTIONS IN THE CONDENSED STATE

Ref. No.	Reactants	Measurements in	Temp. range	No. of temps.	$\log A$ ($A = $ l/mole sec)	E (kcal/mole)	Ref.
1	⎫	Paraffin	−1–172°	7	7.1 ± 0.2	17.9 ± 0.3	178
2	⎪	Pure liquid	−2–80°	10	5.8 ± 0.5	16.2 ± 0.8	179
3	⎪	CCl$_4$	0–55°	5	6.7 ± 0.3	17.1 ± 0.4	178,179
4	⎬ 2 Cyclopentadiene	C$_6$H$_6$	15–55°	6	6.1 ± 0.4	16.4 ± 0.6	143
5	⎪	CS$_2$	0–35°	3	6.2 ± 0.4	16.9 ± 0.6	180
6	⎪	PhNO$_2$	0–55°	4	5.8 ± 0.3	15.1 ± 0.4	180
7	⎪	EtOH	0–55°	4	6.4 ± 0.8	16.4 ± 0.8	180
8	⎭	Dioxane*	40–80°	5	6.9 ± 0.2	17.3 ± 0.4	105
9	⎫	EtOH	2–39°	6	7.0 ± 0.6	12.7	181
10	⎪	n-Hexane	20–50°	4	6.5 ± 0.8	12.1 ± 1.0	182
11	⎪ Cyclopentadiene;	C$_6$H$_6$	8–50°	5	6.5 ± 0.3	11.6 ± 0.7	143
12	⎬ benzoquinone	PhNO$_2$	7–50°	11	6.0 ± 1.0	10.0 ± 1.2	182
13	⎪	PhCN	10–50°	6	4.5 ± 0.2	8.0 ± 0.3	183
14	⎪	CCl$_4$	2–50°	13	4.3 ± 0.3	8.8 ± 0.4	182
15	⎭	CS$_2$	3–39°	7	4.0 ± 0.6	8.5 ± 0.6	182
16	⎫ Cyclopentadiene;	Decalin	0.3–40°	5	7.9 ± 0.4	12.3 ± 0.5	152
17	⎬ maleic anhydride	Dioxane*	15–45°	4	5.4 ± 0.4	8.9 ± 0.5	105
18	⎭	Acetone	−60 to −40°	2	5.0 ± 0.8	8.5 ± 1.0	184
19	⎫ Anthracene;	CHCl$_3$	25–45°	2	5.2 ± 0.8	13.1 ± 1.0	185
20	⎬ maleic anhydride	Dioxane*	120–150°	4	6.7 ± 0.4	16.3 ± 0.6	186

TABLE 14 (Contd.)

Ref. No.	*Reactants*	*Measurements in*	*Temp. Range*	*No. of Temps.*	*$\log_{10} A$ (A=1/mole sec)*	*E (kcal/mole)*	*Ref.*
21	9,10-Dimethylanthracene; maleic anhydride	$CHCl_3$	2–25°	2	4.7 ± 0.6	8.0 ± 0.8	185
22	9,10-Dimethylanthracene; maleic anhydride	Dioxane*	60–100°	3	6.0 ± 0.4	10.9 ± 0.5	186
23	9,10-Dimethylanthracene; maleic anhydride	Acetone	2–25°	2	6.2 ± 0.6	11.3 ± 0.8	185
24	Cyclohexadiene; maleic anhydride	Dioxane*	40–70°	5	5.2 ± 0.4	12.6 ± 0.5	186
25	Butadiene; benzoquinone	C_6H_6	25–55°	4	5.2 ± 0.6	14.5 ± 0.6	187
26	Cyclopentadiene; chloranil	C_6H_6	10–60°	4	6.2 ± 0.4	14.5 ± 0.5	188
27	Cyclopentadiene; α-napthaquinone	C_6H_6	12–51°	3	4.8 ± 0.9	10.0 ± 1.0	143
28	Cyclopentadiene; cyclopentadiene-benzoquinone	C_6H_6	12–55°	5	5.5 ± 0.9	13.2 ± 1.0	143
29	Cyclopentadiene; acrolein	C_6H_6	5–76°	8	6.1 ± 0.3	13.7 ± 0.5	143
30	2-Isobutyl-1:3-butadiene; maleic anhydride	C_6H_6	25–35°	2	1.8 ± 1.3	16.5 ± 1.5	184
31	2-Neopentyl-1:3-butadiene; maleic anhydride	C_6H_6	25–35°	2	2.3 ± 1.3	7.6 ± 1.5	184
32	Hexachlorocyclopentadiene; maleic anhydride	Dioxane	not given		5.8 ± 0.5	19.1 ± 0.6	189

* The figures in column 4 and 5 of this line were given by Dr. J. SAUER in a private communication.

The influence of solvents on the kinetic parameters A and E is shown by some of the data in Table 14. In the dimerization of cyclopentadiene (ref. nos. 1–8) solvent effects are small, but if the reactants contain polar groups considerable variations of kinetic A factors and activation energies occur (ref. nos. 9–15). This indicates that on passing from initial to transition state a change of polarity occurs*. In the addition of benzoquinone to cyclopentadiene there is a linear functional relationship between log A and the activation energy, E, similar to that previously observed in other reactions,[191] and successfully explained[192] with the help of collision theory. An alternative explanation[182] makes use of (52) and (53) between activation energies and heats of solution and between kinetic A values and weight factors W^s

$$E^m - E^n = \triangle H_t^{s(n)} + \triangle H_a^{s(m)} + \triangle H_b^{s(m)} - \triangle H_t^{s(m)} - \triangle H_a^{s(n)} - \triangle H_b^{s(n)} \quad (52)$$

$$\log A^m - \log A^n = \log W_t^{s(n)} + \log W_a^{s(m)} + \log W_b^{s(m)} - \log W_t^{s(m)} - \log W_a^{s(n)} - \log W_b^{s(n)} \quad (53)$$

The subscripts a, b, and t indicate that the relevant quantities relate to diene, dienophile and transition state and superscripts m and n have the same significance as in (41) and (42). There are correlations[193] between entropies and heats of solution, which can be written as follows

$$\log W_a^{s(m)} - \log W_a^{s(n)} = \gamma(\triangle H^{s(m)} - \triangle H^{s(n)}) + q_a \quad (54)$$

with analogous expressions for log W_b^s and log W_t^s. Substituting the terms on the right hand side of (54) and of its analogues for

---

* It is instructive to consider the case of a cyclo-addition occurring in two steps *via* a dipolar intermediate, x, namely

$$a + b \underset{k_1}{\overset{k_2}{\rightleftharpoons}} x \overset{k'_1}{\to} c$$

It follows from the stationary state approximation that
$$dc/dt = k'_1 k_2 [a][b]/(k_1 + k'_1)$$
and it is reasonable to assume that k_2 increases and that both k_1 and k'_1 decrease if one passes from a non-polar to a polar solvent. On the other hand, the ratio $k'_1 k_2/(k_1 + k'_1)$ need not significantly depend on solvent polarity.

the terms on the right hand side of (53) and taking (52) into account, leads to

$$\log A^m - \log A^n = \mathrm{r}(E^m - E^n) + (q_a + q_b - q_t)$$

which is the observed[182] functional relationship between log A and E.

It has been reported[194] that large kinetic solvent effects occur also in the following Diels–Alder reaction

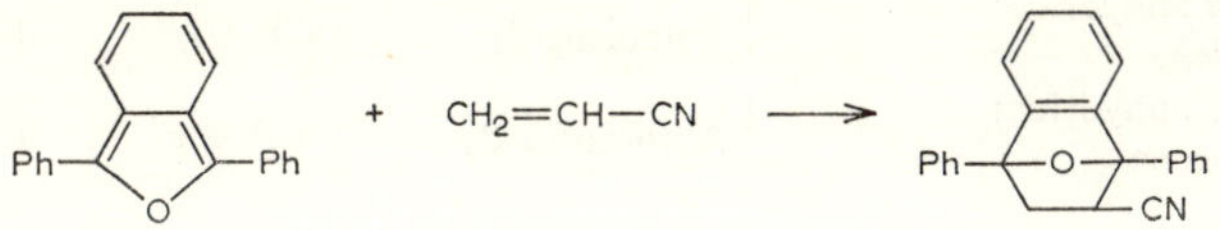

The log A and E values, relating, for instance, to chloroform and carbon tetrachloride are given[194] as 2.2 and 14.5 kcal/mol and 5.7 and 22.4 kcal/mol but such large variations in aprotic solvents of low dielectric constant are unlikely. As details of typical runs (*e.g.* concentrations and temperature) have not been published, it is desirable that this work should be repeated and that experiments be carried out which show that second order reactions give rise to the same adduct in all these solvents.

An investigation into the deuterium isotope effect has been carried out[194a] for three associations in which the dienophile was deuterated and for two reactions in which the deuterium was introduced into the diene. The addenda, solvents and results are shown in Table 14a. These experiments were carried out between 0° and 100°, but the ratios velocity constant of deuterated reactants/velocity constant of protium analogues k_D/k_N, listed in the last column were corrected for 25°. The ratios are not much larger than unity; their mechanistic significance is not clear, for reasons similar to those mentioned below in section 4,3. In particular the conclusion[194a] that "little charge formation is expected at the transition state and that the isotope effect observed should be purely that of hybridisation change" is not supported by other evidence.

Some of the figures in Tables 13 and 14 are of interest because

TABLE 14a
DEUTERIUM ISOTOPE EFFECTS IN DIELS-ALDER REACTIONS

Addenda	Deuterated reactant	Solvent	k_D/k_H
Maleic anhydride; cyclopentadiene	Maleic anhydride-d_2	CH_2Cl_2	1.03
Maleic anhydride; butadiene		CH_2Cl_2	1.007
Maleic anhydride; anthracene		Toluene	1.05
Maleic anhydride; butadiene	Butadiene-d_4	CH_2Cl_2	1.07
Maleic anhydride; anthracene	Anthracene-d_2	Toluene	1.06

they enable a calculation of numerical values of the differences $\triangle_1$ and $\triangle_2$

$$\triangle_1 = \log A \text{ (gas)} - \log A \text{ (benzene solution)} \quad (55)$$
$$\triangle_2 = E \text{ (gas)} - E \text{ (benzene solution)} \quad (56)$$

In the addition of acrolein to cyclopentadiene, $\triangle_1 = 0.1 \pm 0.8$ and $\triangle_2 = 1.5 \pm 1.3$ kcal/mol and in the dimerization of cyclopentadiene, $\triangle_1 = 0.0 \pm 0.8$ and $\triangle_2 = 0.3 \pm 1.2$ kcal/mol. Similar $\triangle$-values can be calculated for other solvents (see ref. nos. 1–8 in Table 14). It has to be taken into account, however, that A (gas) and E (gas) of (55) and (56) were derived from measurements at constant volume, while A(solution) and E(solution) relate to constant pressure. For the dimerisation of cyclopentadiene in the gas phase and in the pure liquid state it is also possible, however, to calculate

$$\triangle_3 = \log A \text{ (gas)} - \log A^v \text{ (condensed state)} \quad (57)$$
$$\triangle_4 = E \text{ (gas)} - E^v \text{ (condensed state)} \quad (58)$$

where the superscript indicates that the kinetic parameters of the reaction in the condensed state relate to constant volume conditions; the last term of (58) is given by

$$E^v\text{(condensed state)} = E^p\text{(condensed state)} + RT^2 \alpha_p \gamma/\beta$$

In this latter expression α_p is the coefficient of thermal expansion

at hydrostatic pressure p, β is the compressibility and γ, defined by
$$\gamma = (\delta \log k/\delta p)_T \qquad (59)$$
represents the influence of hydrostatic pressure on the velocity coefficient k. Determinations[195] of α_p, β and γ enabled one to compute[196] the numerical values in Table 15. The $\triangle_1$ and $\triangle_2$-values are functions of the hydrostatic pressure, p, while $\triangle_3$ and $\triangle_4$ are independent of p up to 3000 atm. It appears, therefore, that A^v and E^v are more characteristic of this liquid phase Diels–Alder reaction than A^p and E^p. The $\triangle_3$ value in column 3 of Table 15 is of interest in connection with theoretical calculations[197] of the bimolecular collision frequency in the condensed state.

In associations of the type here considered, (59) is related to $\triangle V^*$, the difference between the partial molar volumes of initial and transition state by[198]
$$(\delta \ln k/\delta p)_T = -\triangle V^*/RT \qquad (60)$$
The parameter $\triangle V^*$ can be deduced[198] from the slope of plots of $\log k$ versus the hydrostatic pressure p (method 1). If, as a second approximation, the influence of p on $\triangle V^*$ is taken into account, and if it is assumed that the compressibility of the transition state is similar to that of ordinary liquids, one can deduce relationships[199] from which it follows[199] that, within a specified pressure range, plots of $\{\log(k_p/k_0)\} \times p^{-1}$ versus $p^{0.523}$ should be linear. This is in accordance with observations[199]. From such plots $\triangle V^*$ can be calculated by extrapolation to $p = 0$ (method 2). The activation parameter $\triangle V^*$ estimated by these methods for three Diels–Alder reactions are listed in Table 16. The experiments showed that in all cases the rate constants increase with increasing hydrostatic pressure, thereby indicating [198] that a volume contraction occurs if

TABLE 15
ACTIVATION PARAMETERS A^v AND E^v OF THE DIMERISATION OF PURE LIQUID CYCLOPENTADIENE

$-\log A^v$ (condensed state) (A^v in l/mole-sec)	E^v (condensed state) kcal/mole	$\triangle_3$	$\triangle_4$ (kcal/mole)
8.7 ± 0.6	19.8 ± 0.7	-2.6 ± 1.0	-3.1 ± 1.3

TABLE 16
ACTIVATION PARAMETERS, ΔV^*, AND VOLUME CHANGES, ΔV, OF DIELS–ALDER REACTIONS IN THE CONDENSED STATE

Reactants	Measurements in	Temp.	$-\Delta V^*$ (ml/mole) method (1)	$-\Delta V^*$ (ml/mole) method (2)	$-\Delta V$ ml/mole	Ref.
2 Cyclopentadiene	Pure liquid	0°	15	24	31	153a
		20°	18	31	32	195
		30°	22	30	33	
		40°	25	33	33	
	n-Butyl chloride	20°	20.2			153a
		30°	22.3			
		40°	22.5			
2 Isoprene	Pure liquid	60°	24	37	45	199
		70°	25	38	49	200
2,3-Dimethyl-butadiene; butyl acrylate	Bromobenzene	80°	22.6			153a

one passes from the initial to the transition state. The figures in the sixth column of Table 16 show the decrease of volume on passing from initial to final state. The latter contractions are larger, which implies that these transition states are more bulky than the fully formed products.

4. 2 Formation of geometrically isomeric adducts

The data in Table 17, taken from ref. 121, characterize the kinetics of formation of three pairs of endo–exo-isomers (I) and (II) obtained from cyclopentadiene and the three dienophiles

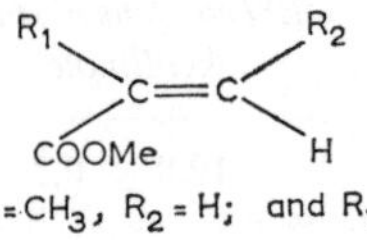

$$\underset{\text{COOMe}}{\overset{R_1}{\underset{}{>}}}C=C\underset{H}{\overset{R_2}{<}}$$

($R_1 = R_2 = H$; $R_1 = CH_3$, $R_2 = H$; and $R_1 = H$, $R_2 = CH_3$)

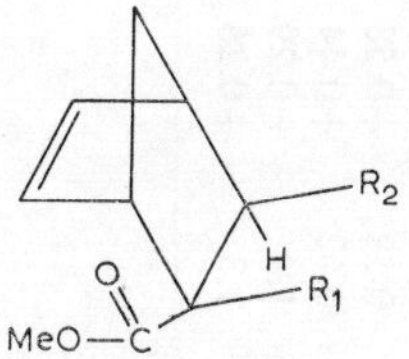

(I) = endo - adduct 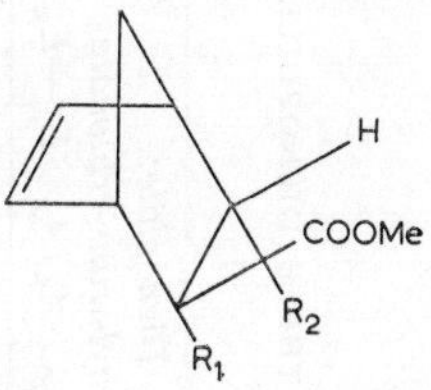   (II) = exo - adduct

The figures in Table 18, calculated from the data in ref. 105 relate to the rate of formation of *cis-trans* isomeric adducts, (III)–(VI), obtained by reactions of *cis-* and *trans-*dienophiles,

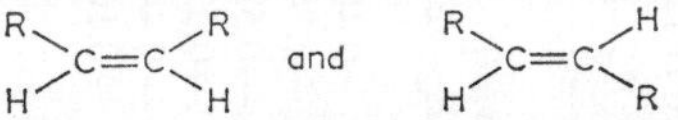

with cyclopentadiene and 9,10-dimethylanthracene.

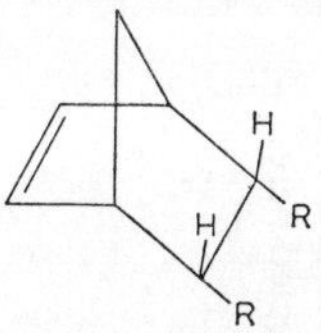

(III) 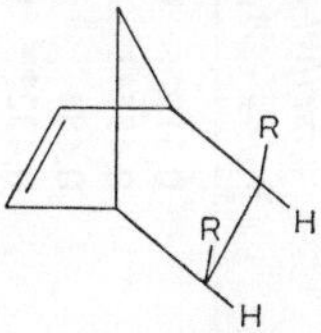   (III') 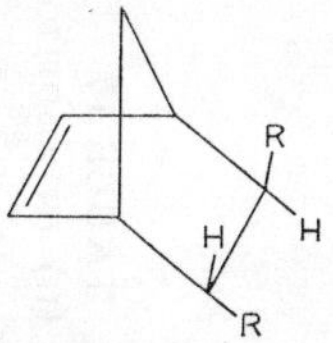   (IV)

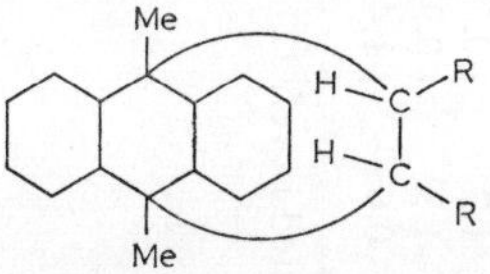

(V) 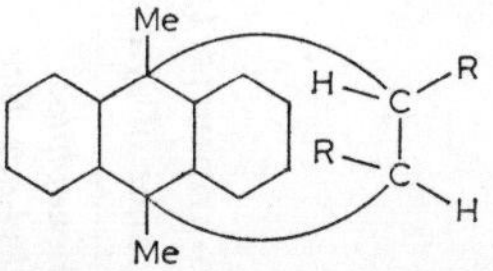   (VI)

Ratios of velocity coefficients and of non-exponential factors in Tables 17 and 18 are designated by k_n/k_x, k_t/k_s, A_n/A_x and A_t/A_s and differences of activation energies by E_x-E_n and E_s-E_t; the subscripts n, x, s and t relate to the formation of respectively endo-exo- and *cis-trans*-isomeric adducts. The ratios k_n/k_x and k_t/k_s depend on differences of activation energies and entropies. Table 18 shows that in some systems the energy difference is of predominant importance, but systems are also known where the

TABLE 17

FORMATION OF ENDO- AND EXO-ADDUCTS (I) AND (II) IN ADDITIONS OF CYCLOPENTADIENE TO THREE DIENOPHILES

For significance of symbols see text. k_n/k_x relates to 30°; $E_x - E_n$ in kcal/mole.

Solvent*	Dienophile: methyl-acrylate			Dienophile: methyl-methacrylate			Dienophile: methyl-trans-crotonate		
	k_n/k_x	A_n/A_x	$E_x - E_n$	k_n/k_x	A_n/A_x	$E_x - E_n$	k_n/k_x	A_n/A_x	$E_x - E_n$
Et$_3$N	2.79	1.2	+0.54	0.319	0.62	−0.40	0.904***	0.30	+0.68
Decalin	3.39**	1.9	+0.35	0.457**	1.6	−0.74	1.09	0.69	+0.34
CH$_3$NO$_3$	4.79	1.7	+0.64	0.484	1.1	−0.52	1.42	0.58	+0.58
Acetone	3.98**	1.2	+0.73	0.432**	0.76	−0.33	1.22	0.62	+0.44
CH$_3$COOH	6.65	1.3	+0.96	0.624	1.3	−0.44	2.33	1.4	+0.29
MeOH	6.84**	0.63	+1.4	0.692**	0.59	+0.10	1.87	1.3	+0.26

* Six out of the twelve solvents listed in ref. 121 are selected here.
** Interpolated values.
*** At 3° $k_n/k_x = 1.05$.

TABLE 18
FORMATION OF CIS-TRANS-ADDUCTS FROM CYCLOPENTADIENE AND 9,10-DIMETHYLANTHRACENE, AND THREE PAIRS OF CIS-TRANS-DIENOPHILES, IN DIOXANE SOLUTION

For significance of symbols see text; $E_s - E_t$ in kcal/mole.

Adducts	Substituent R in cis-trans-dienophile		
	$Me-O-C-$ $\parallel$ O	$Ph-C-O-$ $\parallel$ O	$CN-$
(III) or (III') and (IV)			
k_t/k_s (40°)	82 ± 10	149 ± 20	1.0 ± 0.1
A_t/A_s	0.44 ± 0.03	1.5 ± 0.3	7.0 ± 0.7
$E_s - E_t$	3.3 ± 1.0	2.9 ± 0.6	1.2 ± 0.6
(V) and (VI)			
k_t/k_s (100°)	105 ± 10	—	1.3 ± 0.2
A_t/A_s	0.09 ± 0.01	—	1.0 ± 0.2
$E_s - E_t$	5.6 ± 0.8	—	0.1 ± 0.6

entropy difference plays the major role. In the formation of geometrically isomeric adducts from 2,3-dimethylbutadiene and *trans*- and *cis*-1,2-dibenzoylethylene, for instance, E_t and E_c agree within the limits of the experimental errors, while the ratio A_t/A is about 500 (see ref. 200a).

4. 3 Diels–Alder retrogressions

Table 19, with the accompanying notes, shows the reactions studied, the experimental methods and the activation parameters A and E. The gaseous retrogressions are homogeneous in glass vessels and in the initial stage the reactions are kinetically of the first order. With increasing conversion, the influence of simultaneous, consecutive and reverse processes complicates the overall kinetics, but in the condensed state these complications could be avoided and it could be shown that the reactions are of the first-order up to relatively high conversion.

The activation energies of these retrogressions are in the range 26–57 kcal/mole; they are larger than those of the corresponding forward reactions, the difference being due to the exothermicity of the overall associations. The log A values are in the range 11–14.5; assuming that the transmission coefficient in (51) is unity, entropies of activation, $\triangle S^*$, can be calculated. In the typical case of reaction ref. no. 2, Table 19, $\triangle S^*/R$ is 0.7 ± 0.7, which is small compared to the $\triangle S^*/R$ values of the corresponding forward reaction (see p. 51). The small $\triangle S^*/R$ values of Diels–Alder retrogressions are responsible for the approximate agreement between $\triangle S^*/R$ and $\triangle S/R$ in the case of the combination of dienes with dienophiles.

A secondary isotope effect has been observed[203] in the retrogression of some of the adducts (XIII)–(XVII) to produce methyl-furan and maleic anhydride; the kinetics of these reactions were measured in a mixture of dioxan and isooctane at 50°. The following ratios, r, of velocity coefficients were reported: $r_1 = k_{XIII}/k_{XIV} = 1.16$; $r_2 = k_{XIII}/k_{XV} = 1.08$; $r_3 = k_{XVI}/k_{XVII} = 1.00$.

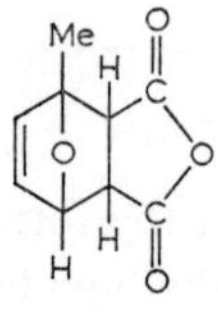

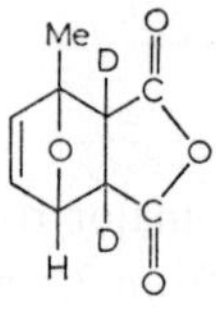

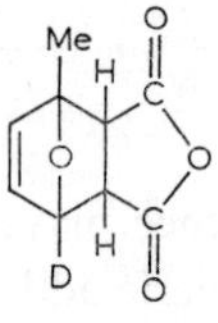

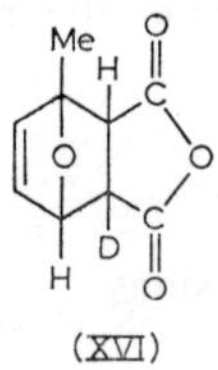

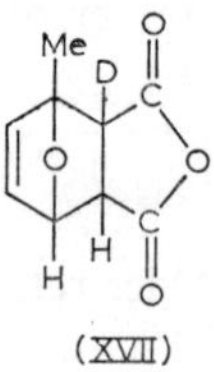

TABLE 19

KINETIC PARAMETERS A AND E OF DIELS–ALDERS RETROGRESSIONS

Ref. no.	Formula of reactant	Products of retrogression	Measurements in	Temp. range	$\log_{10} A$ (sec^{-1})	E $(kcal/mole)$	Technique	Ref.
1	(VII)	Ethylene; butadiene	Gas	485–565°	12.9 ± 1.0	57.5 ± 1.2	a	201
2	(VIII)	2 Cyclopentadiene	Gas / Pure liquid / Paraffin	−1–222° / 100–155° / 120–175°	13.0 ± 0.3 / 13.6 ± 0.4 / 13.0 ± 0.2	33.7 ± 0.5 / 35.3 ± 0.6 / 34.2 ± 0.4	a ; b / c / c	159,176,177 / 202 / 202
3	(IX)	Cyclopentadiene; acrolein	Gas	192–242°	12.3 ± 1.0	33.6 ± 1.2	a	162,177
4	(X)	Cyclopentadiene; benzoquinone		55– 79°	12.6 ± 1.0	29.0 ± 1.2		
5	(XI)	Cyclopentadiene; naphthaquinone	C_6H_6	49– 79°	11.0 ± 1.0	26.0 ± 1.2	c	202
6	(XII)	9-Phenylanthracene; maleic anhydride	o-Dichlorobenzene	132–175°	14.5 ± 0.4	36.9 ± 0.6	d	152

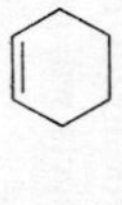

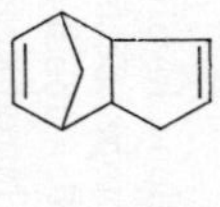

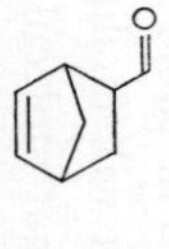

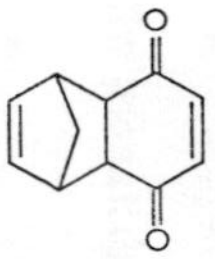

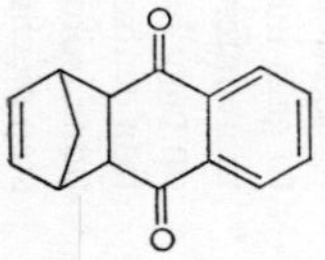

Experimental techniques:

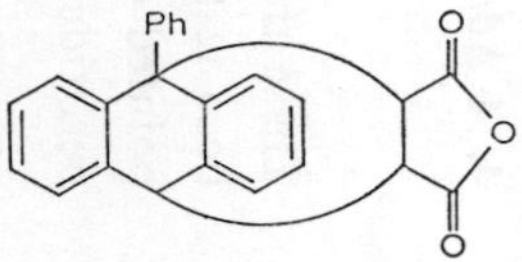

a) Change of initial pressure with time, at constant volume.
b) Determination of equilibrium constant in gas phase (from kinetic measurements in paraffin and solubility determinations and rate of dimerisation of cyclopentadiene in gas phase (see ref. no. 3, Table 13).
c) Estimation of cyclopentadiene which was quickly removed from reaction mixture.
d) Optical density and isotope exchange measurements.

The ratios r_1, r_2, and r_3 are related to the frequency v_1, of the normal modes by[204]

$$r = \left(\frac{m_{\neq}}{m'_{\neq}}\right)^{1/2} \frac{f_{\text{transition state}}}{f_{\text{reactant}}} \cdot e^{\Delta\varepsilon/RT}$$

$$f_{\text{transition state}} = \prod_{3n-7} \left\{\frac{u_1[1-\exp(-u'_1)]}{u'_1[1-\exp(-u_1)]}\right\} \quad (61)$$

$$f_{\text{reactants}} = \prod_{3n-6} \left\{\frac{u_1[1-\exp(-u'_1)]}{u'_1[1-\exp(-u_1)]}\right\}$$

$$u_1 = hv_1/kT$$

where $m_{\neq}$ is the reduced mass for motion along the reaction coordinate, n is the number of atoms and $\Delta\varepsilon$ is the difference between zero point energy of initial and transition states, the primed quantities relating to the deuterated reactants. The formulae show that r_1, r_2, and r_3 depend, inter alia, on unknown v_1 values (particularly those of low frequency). Contrary to the opinion expressed in ref. 203, it is believed, therefore, that the mechanistic significance of this particular secondary isotope effect is not clear*.

4. 4 Added substances without influence on the rate of Diels–Alder reactions

These are listed in Table 20, together with the reactions which have been tested. It will facilitate the discussion in section 5.3, if it is pointed out that molecular oxygen[206], decomposition products of peroxides[206], iodine[206], and possibly also nitric oxide[207] and

---

* In other reactions, involving *e.g.* a solvolysis or acetolysis, a rationalization of secondary isotope effects has been achieved; see ref. 204a and 204b. In order to gain information about the transition state of Diels–Alder reactions it is of more interest to introduce isotopes in positions 1, 4, 5 or 6 of the schemes 1 or 2 on page 1, than to study effects brought about by isotopes adjacent to the reaction centre. It is doubtful, however, whether in conventional Diels–Alder reactions isotope effects will ever enable one to decide between a one step or two step mechanism.

TABLE 20
ADDED SUBSTANCES WITHOUT INFLUENCE ON THE RATE OF DIELS-ALDER ASSOCIATIONS OR RETROGRESSIONS

Added substance	Reaction	Temp.	Ref.
Molecular oxygen	ref. no. 2, table 13	254°	175
	ref. no. 3, table 13	112°	176
	ref. no. 2, table 14	20°	205
	ref. no. 3, table 14	25°	179
	ref. no. 11, table 14	14°	179
Benzoyl peroxide	ref. no. 2, table 14	20° and 40°	195
Dicyclopentadiene peroxide	ref. no. 3, table 14	25°	179
Iodine	ref. no. 11, table 14	25°	181
$FeCl_3$			
$MnCl_2$	ref. no. 9, table 14	20°	181
$CuCl_2$			
Ferric-dipivaloylmethide	Association of 9-phenylanthracene and maleic anhydride in o-dichlorobenzene	132°	152
Molecular hydrogen			
Nitric oxide	ref. no. 1, table 19	485–568°	201
Propylene			

propylene[207] are scavengers of free radicals, while metal ions oxidise[208] or reduce[208] free radicals. The paramagnetic transition metal complex ferri-dipivaloylmethide accelerates singlet triplet transitions[209].

4. 5 Catalysis of thermal Diels-Alder reactions

General

The addition of gaseous benzoquinone to gaseous cyclopentadiene is a heterogeneous reaction in glass vessels[210], which is probably due to the fact that on passing from initial to transition state the polarity increases[210]. Thus the transition state is characterized by a stronger electrostatic field than the addenda taken separately and the addition reaction should be facilitated near a glass surface, which can be regarded as a two-dimensional polar liquid. In other Diels–Alder reactions, however, *e.g.*, those listed in Table 13, the transition states are not much more polar than the initial states and for this reason these reactions occur wholly in the gas phase. Liquid phase Diels–Alder reactions are catalysed by acids[190,211,212,213], phenols[211,212], heavy metal sulphides[214], and Friedel–Crafts catalysts[215]. The role of co-catalysts is not yet known and therefore it is difficult to draw any mechanistic conclusion from experiments with Friedel–Crafts catalysts. This applies also to results[200a,214b] which show that the relative rate of formation of *cis-trans* and position isomeric Diels–Alder adducts are different (a) in absence of catalysts and (b) under catalysis by aluminium chloride[200a] or stannic chloride[214a].

Kinetics of acid and phenol catalysis. The two associations (62) and (63) and the 13 catalysts listed in the third column of Table 21 have been investigated:

cyclopentadiene + benzoquinone →
endo-cyclopentadiene-benzoquinone (X)   (62)
2 cyclopentadiene → endo-dicyclopentadiene (VIII)   (63)

These reactions can be divided into two groups[211], as to whether the catalytic rate is given by (64) or (65),

$$dx/dt = k(a-x)^2 + k_c^{II}z(a-x)^2 \quad (64)$$
$$dx/dt = k(a-x)^2 + k_c^{I}z(a-x) \quad (65)$$

where a and $a-x$ are the equimolar concentrations of diene and dienophile at zero time and time t, z is the concentration of the catalyst, k is the second order rate coefficient of the uncatalysed association and k_c^{II} and k_c^{I} are catalytic velocity coefficients, the superscripts referring to the fact that the catalysis is either of the second or of the first order with respect to the addenda. Typical results for carbon tetrachloride solution* are in Table 21, which shows the catalysts, k_c values and the activation parameters A_c

---

* For other solvents, see ref. 211.

TABLE 21
KINETICS OF ACID AND PHENOL CATALYSIS IN CARBON TETRACHLORIDE

Ref. no.	Formation of	Catalyst	$0.600\,k_c^{II}$	$6 \cdot 10^4 k_c^{I}$	$\log A_c$	E_c	$\log(A/A_c)$	$E-E_c$
1		Cl$_3$CCOOH	4.1 ± 0.5		0.8 ± 0.6	0.0 ± 0.8	3.5 ± 0.9	8.8 ± 1.2
2		Cl$_2$HCCOOH	0.83 ± 0.05		1.0 ± 0.6	3.4 ± 0.8	3.3 ± 0.9	5.4 ± 1.2
3		ClH$_2$CCOOH	0.14 ± 0.01		2.5 ± 0.6	4.3 ± 0.8	1.8 ± 0.9	4.5 ± 1.2
4		CH$_3$COOH	0.012 ± 0.001		3.9 ± 0.9	7.9 ± 1.0	0.4 ± 1.2	0.9 ± 1.4
5	Cyclopentadiene–	C$_{10}$H$_7$SO$_3$H**	0.92 ± 0.09		3.4 ± 0.6	4.5 ± 0.8	0.9 ± 1.1	4.3 ± 1.2
6	benzoquinone (X)	PhOH	0.25 ± 0.01		1.4 ± 0.7	2.5 ± 0.5	2.9 ± 1.0	6.3 ± 1.3
7		o-Cresol	0.12 ± 0.01		2.2 ± 0.7	3.9 ± 0.9	2.1 ± 1.0	4.9 ± 1.3
8		p-Cresol	0.21 ± 0.02		1.3 ± 0.7	2.4 ± 0.9	3.0 ± 1.0	6.4 ± 1.3
9		Quinol	0.17 ± 0.05*					
10		C$_6$H$_2$Cl$_3$OH	0.042 ± 0.007					
11		C$_6$Cl$_5$OH	0.078 ± 0.087					
12		o-C$_6$H$_4$(OH)(NO$_2$)	0.001					
13		Cl$_3$CCOOH		0.40 ± 0.05	−1.1 ± 1.5	5 ± 2	5.2 ± 1.8	12 ± 2
14		Cl$_2$HCCOOH		0.20 ± 0.03				
15	Dicyclopentadiene	ClH$_2$CCOOH		0.10 ± 0.02				
16	(VIII)	H$_3$CCOOH		0.018 ± 0.002				
17		HCl		7 ± 2	−2 ± 2	4 ± 2.5	−8.7 ± 2.3	13 ± 3
18		C$_{10}$H$_7$SO$_3$H**	1.5 ± 0.5		7.6 ± 0.8	16 ± 1	−0.9 ± 1.1	1 ± 1.4

* This value relates to solvent ethanol, 6°.
** C$_{10}$H$_7$SO$_3$H = naphthalene-2-sulphonic acid.

The k_c^{II} and the corresponding A_c values are in l^2 mole^{-2} sec^{-1}; k_c^{I} and the corresponding A_c values are in l mole^{-1} sec^{-1}, while E_c and E are in kcal/mole. The A_c and E_c values were calculated from experiments between −7 and 59° (see ref. 211); the k_c^{II} and k_c^{I} values relate to 25°, except the k_c^{I} values, ref. nos. 9–11, which relate respectively to 26, 21 and 17°. In the experiments ref. nos. 13–18 the acid concentration was below 0.16 M.

and E_c. The last two columns give log A/A_c and $E—E_c$, where A and E are the activation parameters of the uncatalysed associations (see ref. nos. 3 and 14 of Table 14). It will be seen that the catalysis here considered is essentially due to relatively low activation energies rather than to an increase of activation entropy. Under the specified[211] concentration conditions, (64) applies to the experiments ref. nos. 1–12 and 18 of Table 21, while (65) applies to ref. nos. 13–17. If the trichloroacetic acid concentration is relatively high, e.g. 1 M, rather than below 0.16 M as in the experiments of Table 21, the dimerization of cyclopentadiene is overshadowed by the formation of the trichloroacetic acid ester of a dicyclopentyl derivative[216]. Reaction (63) is not catalysed by phenols or bases. A number of proton-acceptors (see ref. 211) retard or inhibit the acid catalysis of both associations. On plotting the logarithm of the k_c values ref. nos. 1–4 and 13–16 of Table 21 against the logarithm of the dissociation constants of the four carboxylic acids, a linear relationship of the Brönsted type is obtained. This can be explained by the assumption that the acid catalysis of these Diels–Alder reactions involves the proton transfer processes specified below. With phenols, on the other hand, the catalytically most active species are least acidic, and on addition of most of these phenols to a carbon tetrachloride solution of benzoquinone the molar light absorption coefficient (470 mμ) increases with increasing k_c value of the phenol. These observations can be accounted for if one assumes[211] intermediates formed by electron transfer from the phenols to benzoquinone. Both proton and electron transfer processes produce ion pairs, designated below by HS^+....B^- and S^-...H...B^- (S = substrate; HB = catalyst). It is suggested that the second substrate molecule is polarized in the electric field of these ion pairs, thereby facilitating the formation of the transition state of the catalysed association. This effect may be partly responsible for the relatively small activation energies, E_c, in Table 21. There is no indication that under the specified conditions[211,212] the two catalysed Diels–Alder reactions are less stereospecific than in the absence of catalysts.

Acid catalysis. In the specified[211] concentration range the following steps are envisaged

$$S + HB \underset{k_{-1}}{\overset{k_1}{\rightleftarrows}} HS^+ \ldots B^- \tag{66}$$

$$HS^+\ldots B^- + S \overset{k_2}{\to} HSS^+\ldots B^- \tag{67}$$

$$HSS^+\ldots B^- \overset{k_3}{\to} SS + HB \tag{68}$$

where S is benzoquinone or cyclopentadiene, HB is an acid, which may be present as a dimer, SS is the adduct (X) or (VIII) and $HS^+\ldots B^-$ or $HSS^+\ldots B^-$ are ion pairs formed by protolytic fission of the proton donor. Application of the stationary state approximation gives:

$$\text{catalytic rate} = k_1 k_2 [S]^2 [HB]/(k_{-1} + k_2[S]) \tag{69}$$

This is compatible with rate equation (64) or (65) according to whether (70) or (71) holds.

$$k_{-1} \gg k_2[S] \tag{70}$$
$$k_2[S] \gg k_{-1} \text{ and, therefore, } k_2 \gg k_1 \tag{71}$$

The observed catalytic velocity coefficients are given by

$$k_c^{II} = k_1 k_2/k_{-1} \tag{72}$$
$$k_c^{I} = k_1 \tag{73}$$

In (62), the proton transfer (66) involves the benzoquinone rather than the cyclopentadiene, the k_1 values relating to benzoquinone being much larger[211] than the k_c^I velocity coefficient, ref. nos. 13–17, in Table 21. In the case of the catalysed dimerisation of cyclopentadiene, (64) holds with naphthalene-2-sulphonic acid, while (65) applies to the other acids. This can be explained if $k_2^\beta/k_{-1}^\beta \ll k_2^\alpha/k_{-1}^\alpha$, where the superscript α refers to the acids ref. nos. 13–17 in Table 21 and the superscript β to naphthalene-2-sulphonic acid. It may be that $k_{-1}^\beta \gg k_{-1}^\alpha$ or that $k_2^\beta \ll k_2^\alpha$; the latter inequality may be due to the fact that the anion, B^-, of the ion pair reacting in (67) is relatively bulky if naphthalene-2-sulphonic acid is the catalyst. The suggested proton transfer mechanism is supported by the observed functional relationships between catalytic velocity coefficients and acid dissociation constants of the catalysts and by the retardation of the overall rate on addition of proton acceptors.

Phenol catalysis. The kinetics are explained with reference to the following steps

$$S + HB \underset{k_{-I}}{\overset{k_I}{\rightleftarrows}} S^- \ldots H \ldots B^+ \tag{74}$$

$$S^- \ldots H \ldots B^+ + S \overset{k_{II}}{\to} SS^- \ldots H \ldots B^+ \tag{75}$$

$$SS^- \ldots H \ldots B^+ \underset{k_{III}}{\to} SS + HB \tag{76}$$

where $S^- \ldots H \ldots B^+$ represents a hydrogen bonded relatively deeply coloured species formed by electron transfer from a phenol, HB, to benzoquinone, S. It could be suggested that cyclopentadiene is the electron acceptor in (62) but this is improbable, because (63) is not catalysed by phenol. It appears that the benzoquinone functions as a proton- and electron-acceptor*, which is comparable with the behaviour of certain polycyclic hydrocarbons. Applying the stationary state approximation to (74)–(76) one obtains an expression similar to (69). The phenol catalysis obeys (64) rather than (65) with

$$k_c^{II} = k_I k_{II}/k_{-I} \tag{76a}$$

the electron transfer step (74) being relatively fast. This mechanism is consistent with the observed increase of light absorption on mixing benzoquinone with catalytically active phenols, and with the fact that an increase of the acidity of the phenol is accompanied by a decrease of the catalytic effect[211].

Activation parameters A_c and E_c. The values listed in columns 6 and 7 of Table 21 can be divided into two groups according to whether they relate to the catalytic velocity coefficients k_c^{II}, represented by (72) or (76a) or to k_c^I given by (73). The A_c and E_c values of the first groups are given by $A_c = A_1 A_2/A_{-1} = BA_2$ and $E_c = E_{-1} - E_1 - E_2 \sim \triangle H - E_2$ where B is a measure of the statistical weight of the associated species formed in (66) or (74),

---

* The proton acceptor properties of benzoquinone in benzene were established by indicator studies (unpublished).

and $\triangle H$ is the heat change of these equilibria. It has been observed [211] that there is then a functional relationship between log A_c and E_c which may be due to the fact that in these reactions A_2 and E_2 change less markedly than B and $\triangle H$, so that the overall effect is essentially reduced to a correlation between entropy and heat changes[193,198,182].

The A_c values, ref. nos. 13 and 17 in Tables 21, characterizing the proton transfer (66) from trichloroacetic or hydrochloric acid to cyclopentadiene are much smaller than the A values in Tables 13 or 14. It is suggested that the ion pairs in (66) are formed via ionic transition states, which are more strongly solvated than catalyst and substrate taken separately, and that the increased solvation is accompanied by an unfavourable entropy of activation, which accounts for these small A_c values.

4.6 Photo-associations

It has been observed[217] that energy can be transferred from the triplet states of irradiated carbonyl compounds to a conjugated diene (1,3-pentadiene), thereby producing a triplet state of the latter species. It is probable that excited singlet states formed by light absorption of simple conjugated dienes do not decay via triplet levels and it appears, therefore, that the photo-sensitized production of triplet states of conjugated dienes opens up a new field of preparative chemistry. From a theoretical point of view, it is of interest that an estimation of the energy of the diene triplet state is possible, since it is unlikely that energy transfer will occur if the diene triplet lies much above the triplet of the sensitizer. An investigation of chemical reactions of the triplet states of butadiene[218], 2,3-dimethylbutadiene[219], isoprene[219] and maleic anhydride[220] was carried out using as photosensitizers benzil, 2,3-pentanedione, acetophenone, benzophenone, 2-acetonaphthone, benzanthrone, 9-fluorenone, 1-naphthaldehyde, anthraquinone and other carbonyl compounds. Butadiene triplets produced *cis*- and *trans*-1,2-divinylcyclobutane, together with the

Diels–Alder adduct, 4-vinylcyclohexene. Isoprene triplets gave rise to seven dimers, two of which are Diels–Alder adducts:

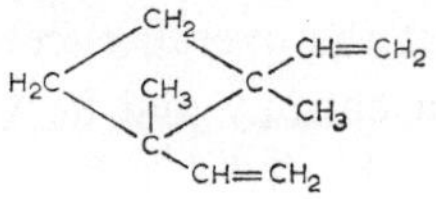

(XVIII) 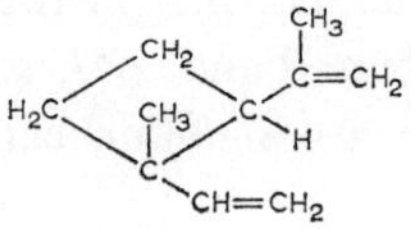 (XIX) 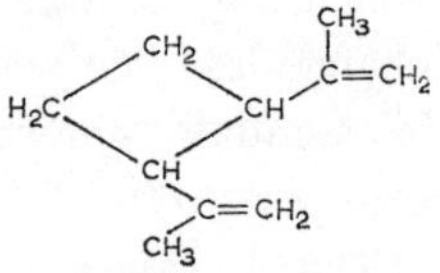 (XX)

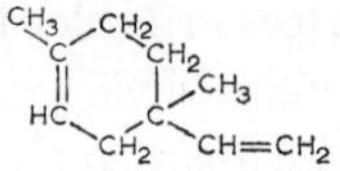

(XXI) 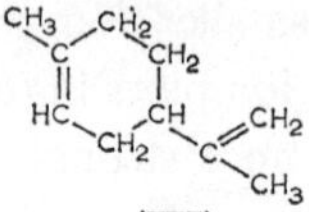 (XXII)

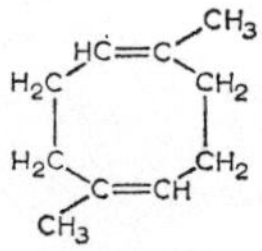

(XXIII) 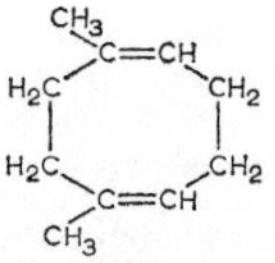 (XXIV)

The composition of the mixture of these dimers seems to depend on the energy requirement to produce the lowest triplet state of the carbonyl compound used as a sensitizer. It is assumed[219] that high energy sensitizers produce predominantly triplets of isoprene in the *trans* configuration, which play a role in the formation of the cyclobutane and cyclooctadiene dimers (XVIII)–(XX) and (XXIII), (XXIV). It is postulated[219] that sensitizers of excitation energy below 60 kcal/mole produce a relatively large proportion of *cis*-triplets of isoprene, which in turn give rise to the Diels–Alder dimers (XXI) and (XXII). The concept of *cis-trans* isomeric triplets of conjugated dienes is new, and attempts have been made to support this hypothesis by a study of the phosphorescence spectrum of ketone (XXV)

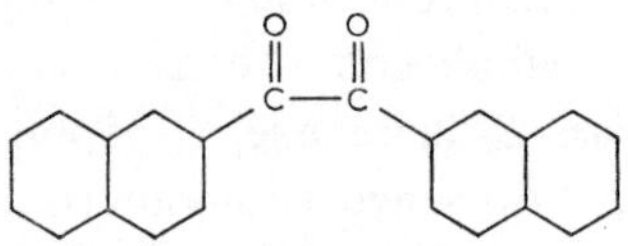

(XXV)

which shows emissions[221] arising from two triplets having different radiation life-times. It has been suggested that the two emitting species are *cis-trans* isomeric triplets of (XXV).

Maleic anhydride triplets react with benzene to from a di-adduct[222]

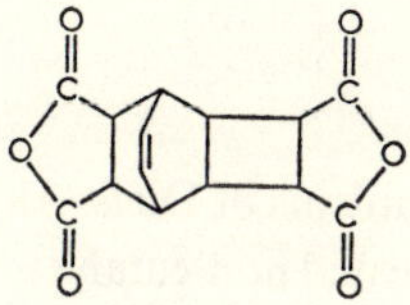

Quantum yield measurements made with different sensitizers indicate[220] that the inefficiency in this reaction arises after formation of maleic anhydride triplets, owing to non-radiative decay.

Chapter 5

MECHANISM

In this chapter two mechanisms of Diels–Alder additions are discussed. The first relates to thermal non-catalysed associations and involves a cyclic non-planar transition state formed in one step from the addenda. Evidence compatible with this mechanism is discussed in section 1 and 2. The second type of reaction involves intermediates in which only one bond between diene and dienophile is fully formed. It will be shown in sections 3–6 that this hypothesis is unsatisfactory for thermal uncatalysed diene syntheses, but it accounts for certain observations in the case of catalysed associations.

5. 1 One-step mechanism; activation energy

(a) Description of the transition state. Fig. 1 represents a cross section[223] through two discontinuous energy surfaces relating to the reactants, R, and to the fully formed product, P, the intersection point, T′, showing, as a first approximation, the energy barrier separating the initial and the final state. Section R–R′ indicates the distortion energy needed to decrease the normal distance between atoms 1 and 4 in the diene and to increase the normal distance between atoms 5 and 6, of the dienophile, the numbering of atoms being as shown below (*cf.* also schemes 1 and 2 on p. 1).

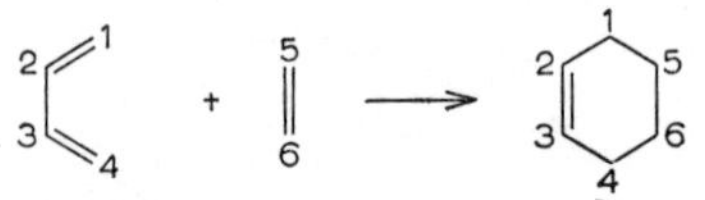

This reaction scheme has to be suitably modified in order to apply it to other types of addenda. If anthracene, for instance, functions as a diene, carbon atoms 9 and 10 (see (49), p. 44) fulfil the role of atoms 1 and 4 in an open chain diene. The second section

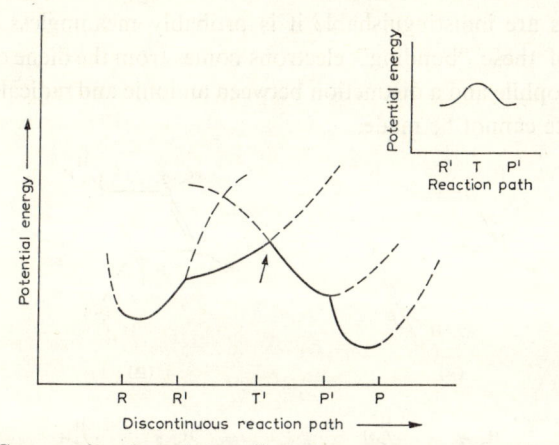

Fig. 1. Cross section through two discontinuous energy surfaces representing the potential energy of reactants, R, and product, P, in a Diels–Alder association (scheme (1), p. 1). The stabilization of the transition state is shown in the inlet.

R'...T' in Fig. 1 represents the energy of interaction between the distorted reactants along a path compatible with stereochemical requirements; the third section, T'...P', corresponds to changes of length of bond distances in the direction of the reaction coordinate and the final section P'...P takes care of bond distances in other directions and of changes in bond angles. These latter energy terms are not incompatible with the assumption, that, on the whole, the nuclear geometries of transition and final state are similar. At the intersection point, T', a switch of electrons occurs and the system changes from the addenda to the adduct. As a second approximation one has to consider stabilization of the transition state by resonance energy[3,223,224] and by non-bonding attraction[109,110,225], the transition state being described[3,223,224] as a hybrid of structures of the type shown in Fig. 2–8. For stereochemical reasons atoms 1 and 4 must be in *cis*-orientation and atoms 1, 4, 5 and 6 cannot be in one plane. The formation of the two incipient bonds is brought about by overlap of molecular π orbitals in a direction corresponding to end-wise rather than lateral overlap of atomic p-orbitals. Each reactant contributes two electrons for the formation of these two incipient bonds. Since electrons are indistinguishable it is probably meaningless to ask which of these "bonding" electrons comes from the diene or from the dienophile and a distinction between an ionic and radical transition state cannot be made.

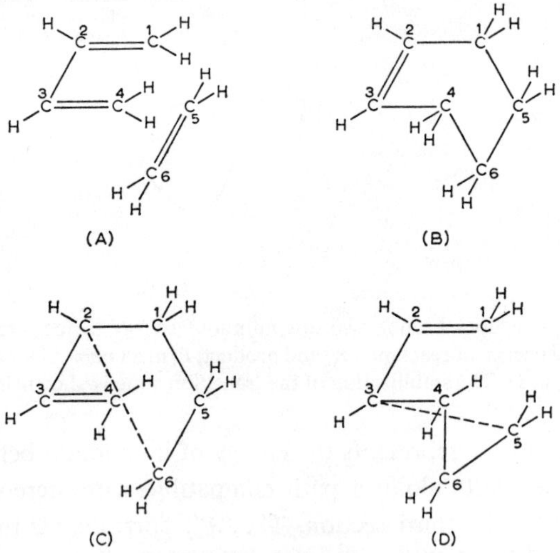

Fig. 2. Structures representing the transition state of the reaction butadiene + ethylene → cyclohexane.

Many Diels–Alder reactions are exothermic (see Table 8). In view of discussions[226, 226a] relating to other exothermic processes it is possible[4] that there is a greater electronic resemblance between initial and transition state than between transition and final state, notwithstanding the similar nuclear geometries of transition and final state[4]. In this situation, structures A in Fig. 2–8 may have a relatively large weight. The distinction between full and dotted lines in the structures of Fig. 2–8 is arbitrary, but the dotted lines are longer and represent weaker bonds, which may even be "formal",

TRANSITION STATE

(A)

(B)

(C)

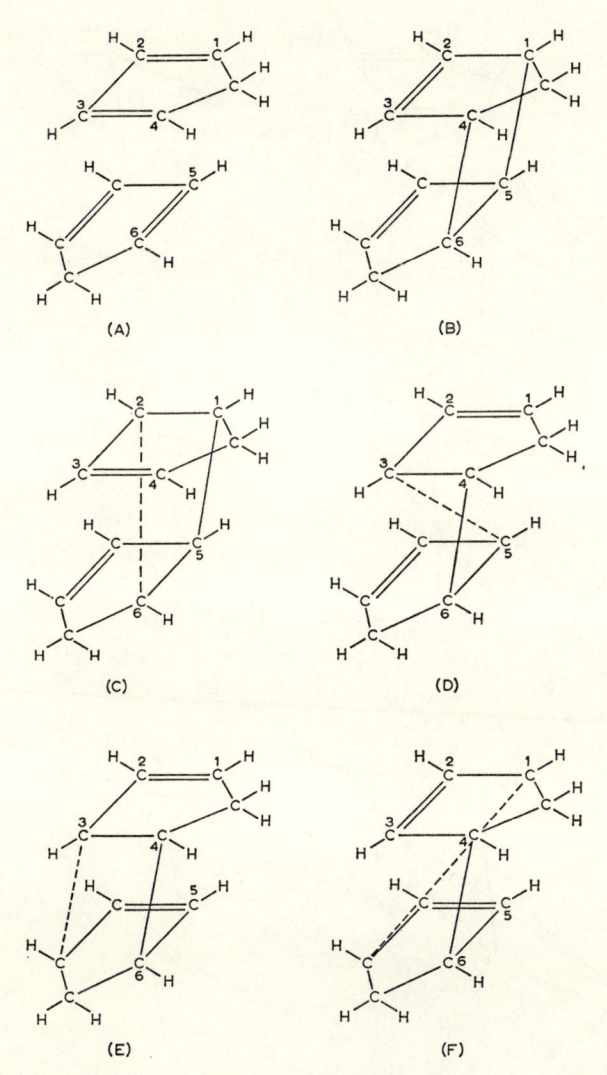

(D)

(E)

(F)

Fig. 3. Structures representing the transition state of the reaction
2 cyclopentadiene → endo-adduct.

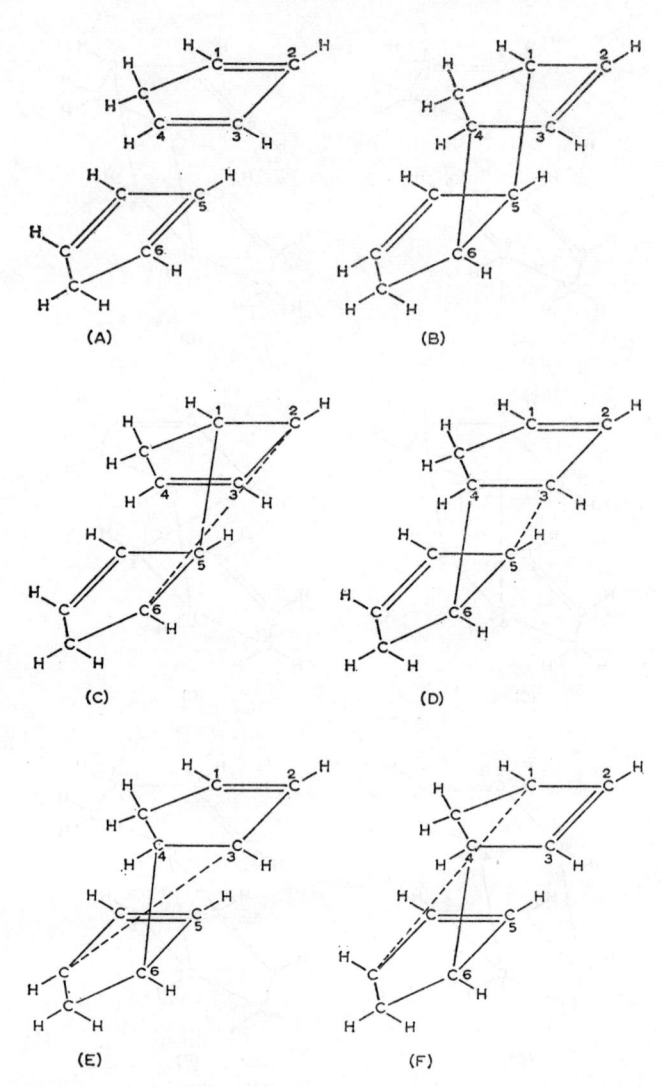

Fig. 4. Structures representing the transition state of the reaction
2 cyclopentadiene → exo-adduct.

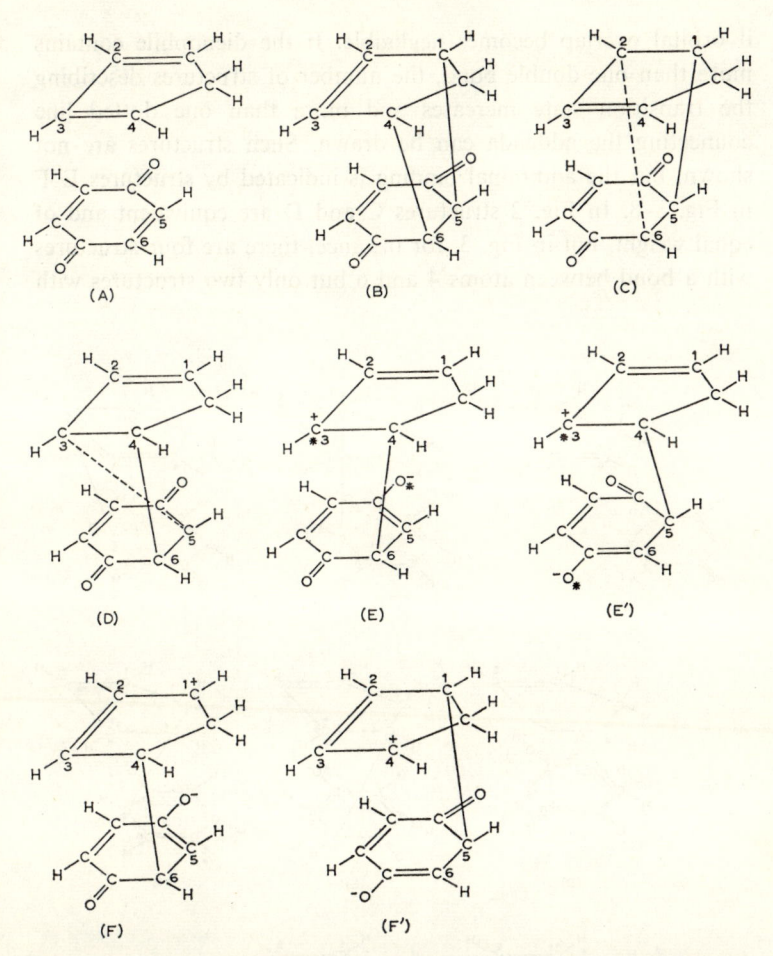

Fig. 5. Structures representing the transition state of the reaction cyclopentadiene + benzoquinone → endo-adduct. The significance of the asterisks in structures E and E' of this Figure and of Figs. 6–8 is discussed in the text (section d).

if orbital overlap becomes negligible. If the dienophile contains more than one double bond, the number of structures describing the transition state increases and more than one dotted line connecting the addenda can be drawn. Such structures are not shown, but the additional binding is indicated by structures E–F in Fig. 3–8. In Fig. 2 structures C and D are equivalent and of equal weight, but in Fig. 3, for instance, there are four structures with a bond between atoms 4 and 6 but only two structures with

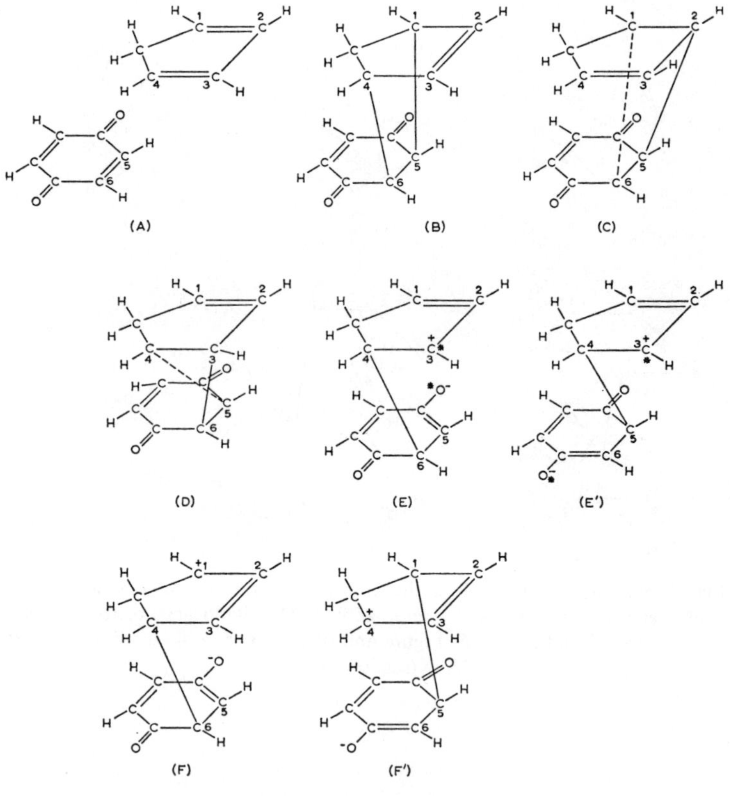

Fig. 6. Structures representing the transition state of the reaction cyclopentadiene + benzoquinone → exo-adduct.

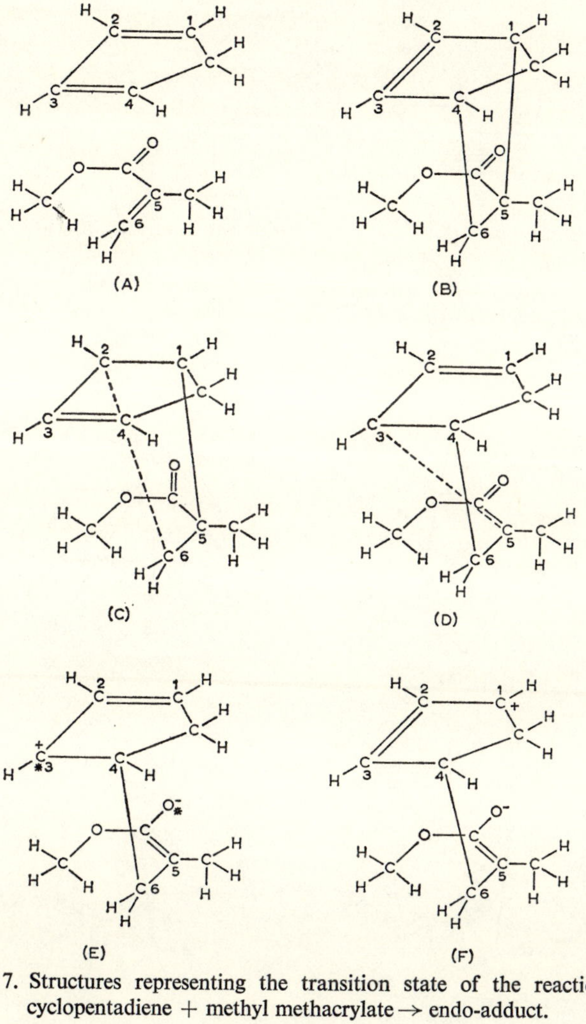

Fig. 7. Structures representing the transition state of the reaction cyclopentadiene + methyl methacrylate → endo-adduct.

a bond between atoms 1 and 5; a similar situation[224] exists in the transition state of the dimerisation of butadiene to produce vinylcyclohexene. In general, therefore, the formalism adopted here does not imply that the two incipient bonds between the atoms 1,4 of the diene and the atoms 5,6 of the dienophile are of equal

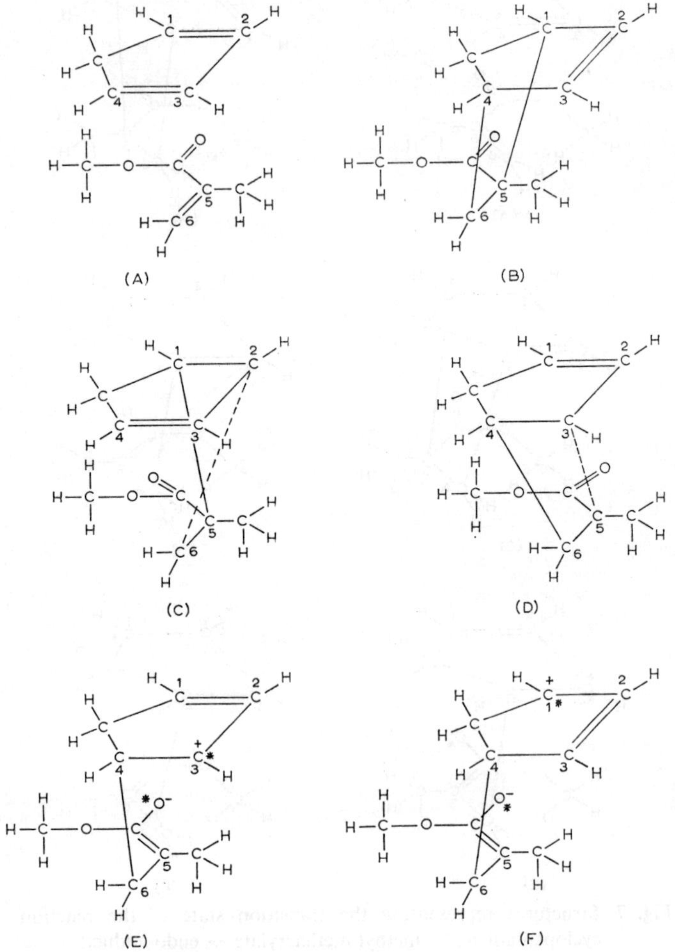

Fig. 8. Structures representing the transition state of the reaction cyclopentadiene + methyl methacrylate → exo-adduct.

strength. In reactions between hydrocarbons, ionic structures are not important, the full and dotted lines in the structures of Fig. 2–8 representing also the contributions of ionic bonds, with two electrons at one end and a vacant orbital at the other end. If one or both reacting species contain polar groups some bonds in the transition state will be polar, as shown in Fig. 5–8, where the dienophile can be regarded as electron acceptor. A similar situation is characteristic of many other dienophiles, but cases are known (see below), where the polarity is reversed.

For purpose of comparison with experiments the above conclusions can be summarized as follows. The activation energy of Diels–Alder reactions will be increased by large distortion energies and by large repulsion (including steric hindrance) between the addenda. Effects which make for a decrease of the activation energies are resonance energy of the transition state and non-bonding attractions, *e.g.* dipole induction or dispersion forces. An attempt will be made in the following sections, (b)–(g), to support these conclusions and to justify the postulated nuclear geometry of the transition state.

(b) Distortion energy. Maleic anhydride reacts with cyclopentadiene and with cyclohexadiene to produce the adducts (I) and (II), the activation energy, E, of the former association being 3.7 ± 1 kcal/mol smaller than the E value relating to the cyclohexadiene adduct (see Table 14, ref. nos. 17 and 24).

The distance between atoms 1 and 4 in normal cyclopentadiene is about 20% smaller than that between atoms 1 and 4 in normal cyclohexadiene. The distortion energy needed to build up the maleic anhydride cyclohexadiene transition state must be larger, therefore, and it is suggested that this accounts for the observed difference between the activation energies.

(c) Resonance energy. The end-on and broadside-on method of addition of a dienophile to a diene leading to linear and cyclic products involves bimolecular collision in which there is close contact between respectively two and four carbon atoms. For this reason the repulsion energy in the formation of the cyclic transition state must be relatively large, the adverse effect being made good, to a certain extent, by the influence of resonance energy[3,223,224]. The linear transition state of the reaction between butadiene and ethylene, for instance can be described by the three structures

$$CH_2=CH-CH=CH_2 \quad CH_2=CH_2$$
$$H_2C=CH-\overset{\bullet}{C}H-CH_2-CH_2-\overset{\bullet}{C}H_2$$
$$H_2\overset{\bullet}{C}-CH=CH-CH_2-CH_2-\overset{\bullet}{C}H_2$$

In the cyclic transition state, on the other hand, the four structures of Fig. 2 play a role and it is reasonable, therefore, to postulate that the latter species is more effectively stabilised. The transition states in Fig. 2, 3 and 5 are described by four, six, and eight structures and it can be assumed, therefore, that resonance energy is at least partly responsible for the trend of the activation energies ref. nos. 1 and 3 in Table 13 and ref. nos. 1–8 and 9–15 in Table 14. Again, maleic anhydride and isosafrol appear to produce a transient Diels–Alder adduct[33] (III), while maleic anhydride and styrene or β-methylstyrene do not react in a similar manner, under the same experimental conditions.

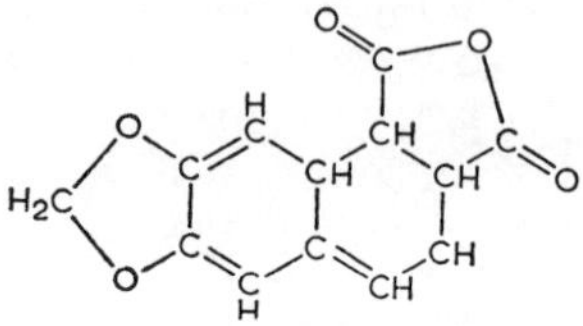

The difference of reactivity has been explained[3] by increased stability of the isosafrole transition state, owing to the contribution of ionic structures of type

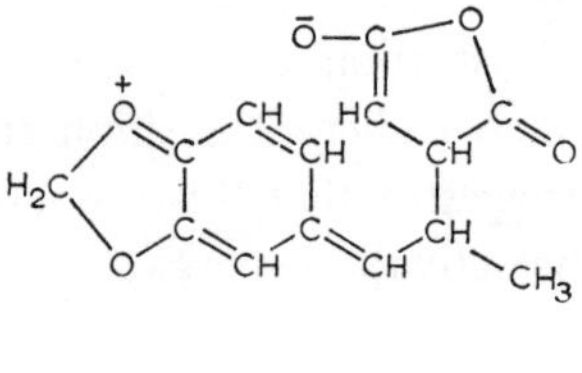

Ionic structures have also been invoked[224] to explain the predominant formation of certain positional isomers discussed in section 2,2. Typical examples[224] are the formation of o- and p-adducts from acrolein and 1-diethylamino- and 3-methoxybutadiene respectively.

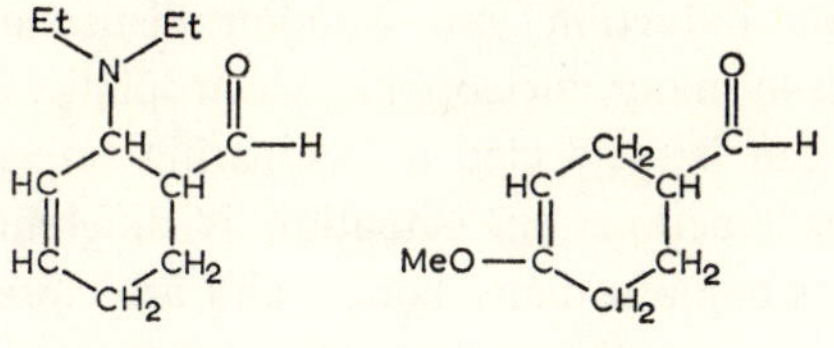

o - adduct   p - adduct

It is assumed that the relevant transition states are stabilised by contributions from

and

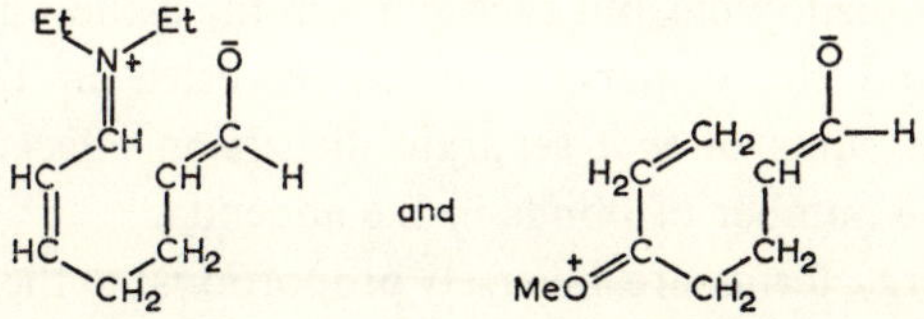

A similar stabilization could not occur if the reaction with diethylaminobutadiene led to a p-adduct and if the addition to the methoxybutadiene led to an o-product.*

In these reactions the dienes can be regarded as electron donors and the dienophiles as acceptors. It has been found[227] that tetrachloro-o-benzoquinone functions with certain reagents as a diene, rather than a dienophile. With styrene, for instance, adduct (IV) is formed and it has been suggested[224] that tetrachloro-o-benzoquinone acts as an acceptor and that the transition state of this reaction is stabilized by contributions from structures of type (V).

---

* It is feasible that in certain systems (see, e.g. ref. 139a) the relative yields of position isomers are also influenced by steric effects; detailed kinetic tests would be required to prove this.

(IV)   (V)

(d) Non-bonding attraction; formation of geometrically isomeric adducts. Dipole induction and London dispersion effects are relatively small in many nucleophilic, electrophilic, or free radical reactions and can be neglected in comparing activation energies. In Diels–Alder reactions the situation is different because the transition states contain many bonds, and are closely packed. In this situation non-bonding attractions contribute substantially to the stability of the transition states and influence the mutual orientation of the addenda, as shown by Table 22 which relates to the addition of cyclopentadiene to benzoquinone. In calculating the dipole induction energies[109,110] the interaction between each dipole in the dienophile and each bond in the diene was separately considered and the dispersion effects[225], listed in the last two columns are sums of $m \cdot n$ separate dispersion effects*, where m and n are the number of bonds in the addenda.

These energy terms are inversely proportional to the sixth power

TABLE 22

NON-BONDING ATTRACTION IN THE TRANSITION STATE OF THE DIELS–ALDER ADDITION OF CYCLOPENTADIENE TO BENZOQUINONE

Distance in Å between carbons 1, 5 and 4, 6 (see Figs. 6 and 7)	Dipole induction (kcal/mole) Orientation		Dispersion (kcal/mole) Orientation	
	endo	exo	endo	exo
2.33	−3.3	−2.5	−25.1	−21.9
2.00	−6.7	−4.0	—	—
1.80	−9.6	−5.3	—	—

* A calculation of the dispersion effect between molecules containing conjugated double bonds has been described by COULSON and DAVIES[228]. This method of calculation is better than that of ref. 225. Semi-empirical functions for the calculation of non-bonding interaction, consisting of an attractive and repulsive term, have been proposed by a number of authors. For literature citations see ref. 228a.

of the distance, d, between the relevant bonds, d being a more important variable than bond polarizability*. The accuracy of the figures in Table 22 is low, but the results show nevertheless that non-bonding attraction contributes substantially to the stability and that the endo-orientation is favoured. In this stereospecific Diels–Alder reaction, the influence of the non-bonding and bonding attraction reinforce each other as indicated by Table 22 and by Figs. 5 and 6: the bond between carbons marked by asterisks in structures E and E' is shorter in the endo- than in the exo-orientation, and thus bonding in the endo-transition state is assumed to be stronger.

Non-bonding attraction in transition states are also of importance for the stereochemistry of Diels–Alder adducts containing more than two fused rings. As an example we consider (19) on p. 32, the product of which can exist in the form of the four isomers shown there. Mutual orientation of the addenda suitable for the formation of these isomers are in Fig. 9. Orientations α and γ are similar to exo-transition states of more simple associations, while β and δ correspond to endo-transition states, orientation β being conducive to the formation of the one isomer which has actually been isolated. Inspection of space filling molecular models shows that, on the whole, transition state β is more closely packed than the other three transition states. To see this, the distances between all the atoms, also those forming the methylene bridge must be taken into account. These space filling models do not reveal that steric hindrance operates in the formation of any of the transition states. It appears, on the other hand, that bonding and non-bonding attractions are strongest if the addenda are in orientation β, and therefore it is reasonable to assume that this is the explanation for the observed stereospecificity of the association.

There are Diels–Alder reactions in which the contributions of bonding and non-bonding attractions appear to oppose each other.

---

* Treatments based on the assumption that the only interactions of importance are those between strongly polarizable multiple bonds (see, for instance, ref. 130) are of doubtful value.

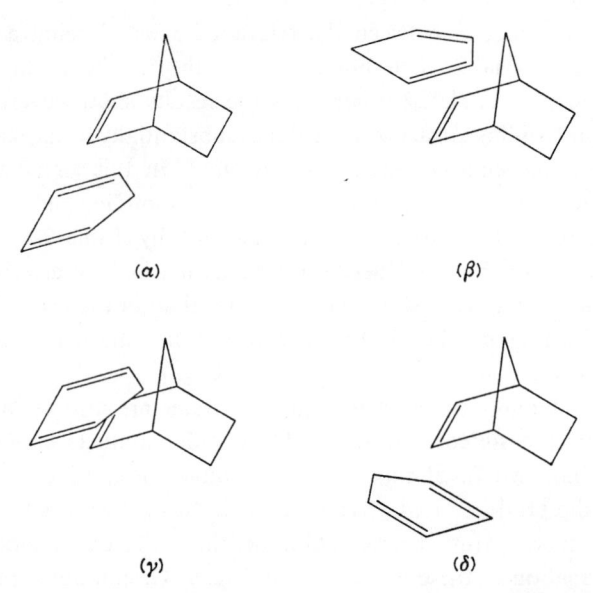

Fig. 9. Mutual orientation of reactants in four transition states suitable for the formation of four geometrically isomeric adducts in the reaction.

The structures α, β, γ and δ correspond to structure A in the preceding figures: other structures contributing to these four transition states are not shown.
Orientation β is conducive to the formation of the observed product.

An example is the addition of methylmethacrylate to cyclopentadiene, transition states suitable for endo- and exo-addition being shown in Figs. 7 and 8. The bond between atoms marked by asterisks in structure E is shorter in the endo-orientation, while the distance, d, between the methylene group of the diene and the ester grouping of the dienophile is markedly shorter in the exo-orientation. It follows that bonding attraction may favour the formation of the endo-adduct, while the non-bonding attraction, owing to the different d-values, could give rise to an increased stability of the exo-transition state. It is suggested that opposing effects of this kind are responsible for the fact that there is no stereospecificity in the reactions referred to in Table 17 and that the differences between activation energies of endo- and exo-addition are so small that the product ratios depend on temperature, solvent and minor structural changes of the reactants. The nature of the addenda influences also the relative rates of reactions between dienes and *cis-trans*-isomeric dienophiles, as shown by the results in Tables 5 and 18. If entropy effects are not of predominant importance it can be concluded that also in these systems the relative stability of transition states suitable for the formation of geometrically isomeric adducts depends markedly on both bonding and non-bonding attraction*. To support these tentative suggestions it would be desirable to carry out calculations similar to those described in ref. 109, 110 and 225.

(e) Dipole–dipole repulsion. If one of the reactants of a Diels–Alder association is a hydrocarbon, the only significant non-bonding interactions are dipole induction and dispersion effects, which stabilize the transition state. If, on the other hand, both addenda are polar molecules, dipole–dipole repulsion might increase the activation energy, E, an effect which could be partly responsible for the relatively high E value of the reaction between hexachloro-cyclopentadiene and maleic anhydride (see ref. nos. 17 and 32 in Table 14).

(f) Conformation of diene. In the case of 1,3-butadiene, the position of the equilibrium (3) on p. 2 between transoid and cisoid conformations is far on the side of the former[102], provided the temperature is not too high[102]. In this situation a Diels-Alder addition to a dienophile should be preceded by conversion of the transoid diene into the *cis* form and the enthalpy change, ΔH, of equili-

---

* Tests with space filling molecular models show that even relatively bulky substituents, *e.g.* $-\overset{\overset{\text{O}}{\|}}{\text{C}}-$OMe, of *cis*-dienophiles do not seriously interfere with space requirements in passing from the initial to the transition state. It is improbable, therefore, that the $E_s - E_t$ values listed in Table 18 are mainly due to steric hindrance in *cis*-dienophiles. The absence of steric hindrance is also indicated by situation (a), referred to in Table 5.

brium (3) should be part of the apparent activation energy of a reaction involving a transoid diene. To test this concept, kinetic comparisons have been made[187] one of which is as follows:

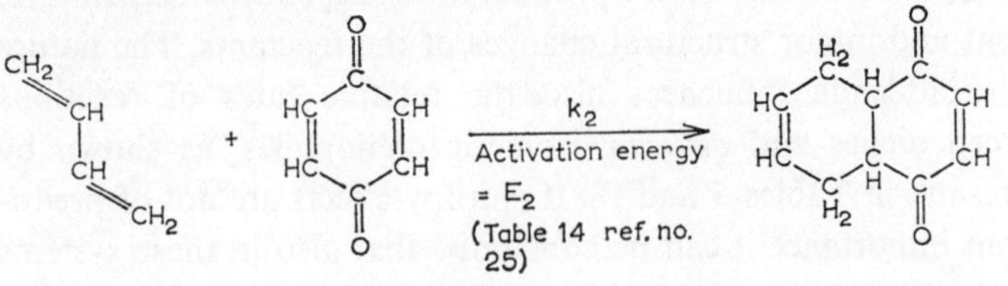

$$\text{butadiene} + \text{benzoquinone} \xrightarrow[\substack{\text{Activation energy} \\ = E_2 \\ \text{(Table 14 ref. no. 25)}}]{k_2} \text{adduct} \qquad (77)$$

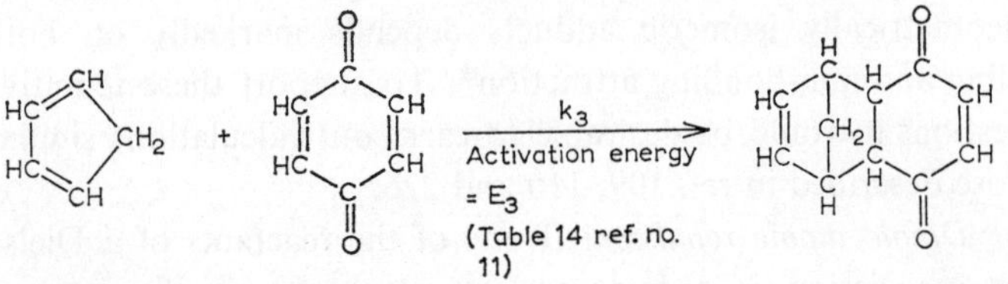

$$\text{cyclopentadiene} + \text{benzoquinone} \xrightarrow[\substack{\text{Activation energy} \\ = E_3 \\ \text{(Table 14 ref. no. 11)}}]{k_3} \text{adduct} \qquad (78)$$

$$\text{cis-butadiene} + \text{benzoquinone} \xrightarrow[\substack{\text{Activation energy} \\ = E_4}]{k_4} \text{adduct} \qquad (79)$$

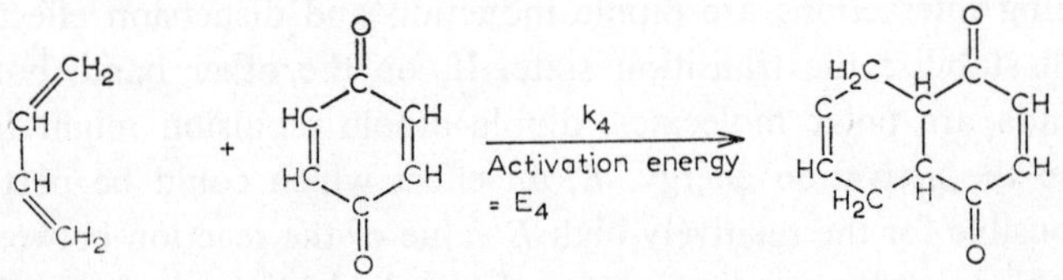

Taking *cis*-butadiene as an intermediate of short life time and applying to (77) and (79) the stationary state approximation with $k_t \gg k_4$ [Benzoquinone] leads to

$$k_2 = k_4 k_c / k_t \text{ and } E_2 = E_4 + \triangle H$$

where k_c and k_t relate to equilibrium (3) on p. 2.

The numerical value of $\triangle H$ is taken[102] as 2.3 kcal/mole, whence $E_4 = 12.2 \pm 0.6$ kcal/mol. It will be seen that E_3 and E_4 agree with each other, within the limits of experimental error specified in Table 14; this is compatible with a one step addition, because (78) and (79) involve both *cis*-dienes and are thus fundamentally similar.

(g) Non-planar transition states. These are formed if a diene and a dienophile approach each other in different planes as shown, for typical associations, in Figs. 2–8. Experimental evidence for

this stereochemical situation is obtained by results of kinetic measurements relating to Diels–Alder reactions of the type

$$\text{(80)}$$

where the substituents X are (a) hydrogen and (b) chlorine. If the six π electron centres were in one plane in the transition state, interacting as in benzene, steric effects due to juxtaposition of the atoms of the diene and the substituent X of the dienophile should be larger in case (b) than in reactions of type (a). The activation energies, E, and the Arrhenius A factors ref. nos 4, 11, 27, 28, 29 and 26 in Table 14 show, however, that, on the whole, there is no marked dependence on the nature of the substituent X. Another series of reactions of interest is the addition of maleic anhydride to anthracene or its derivatives. If the formation of the adducts formed in (49) on p. 44 or of similar species involved planar transition states, large steric effects must operate and the A and E values of the additions of dienophiles to anthracene should be quite different from the activation parameters of more simple associations, *e.g.* (80) with X being hydrogen. The results, ref. nos. 19–23 in Table 14 show, however, that this is not the case.

If the transition states of Diels–Alder reactions were "pseudo-aromatic" planar rings, one could calculate the electronic energy, ε, by a method formally similar to that used for estimating the resonance energy of benzene. Such computations have been carried out[223] and it has been suggested[223] that the ε value of planar transition states is responsible for the relatively small activation energies of Diels–Alder associations. As a standard of comparison overall activation energies were cited[223] which relate to some of the reactions listed in Table 12. If, however, the comments on these reactions on p. 49 are taken into account, the argument in favour of pseudo-aromatic Diels–Alder transition states loses its appeal.

5. 2 One-step mechanism: entropy of activation

It has been pointed out in section 4,1 that the entropies of activation, ΔS^*, derived from the kinetic A values in Tables 13 or 14 are negative. To see whether this can be explained by a loss of translational and rotational degrees of freedom in passing from initial to transition state, without taking care of vibrational contributions, we represent the ΔS^* value of a gaseous Diels–Alder reaction by [168b]:

$$e^{\Delta S^*/R} = \frac{N}{1000} \times \frac{\lambda_t}{\lambda_a \times \lambda_b} \times \frac{r_t}{r_a \times r_b} \varrho^* = W^* \times \varrho^* \qquad (81)$$

where λ and r are translational and rotational partition functions, the subscripts t, a and b indicating here and below that the relevant quantities relate to the transition state and to the reactants. The significance of ϱ^* in (81) is similar to that of ϱ' in (36), p. 40, both factors being functions of vibrational partition functions, ϱ^* and ϱ', relating respectively to a + b $\rightleftharpoons$ transition state and to a + b $\rightleftharpoons$ product. Numerical values of ϱ^* can be estimated[162,163,229] from activation parameters in a manner similar to that employed in the computation of the ϱ' values listed in Table 10. It can thus be established that the ϱ^* factors of the reactions ref. nos. 1, 2 and 3 of Table 13 are of the same order of magnitude as the thermodynamic ϱ' values. There must be vibrational modes of relatively low frequency, both in these transition states and in fully formed adducts which make the entropies of activation and the thermodynamic entropies less unfavourable than one would expect for rigid reactants of comparable size and symmetry. For this reason it has been suggested in section 4,1 that some Diels–Alder associations are "fast" reactions from the point of view of the transition state theory. This consideration shows moreover, that it is quite reasonable to postulate a certain similarity between the nuclear geometries of Diels–Alder transition states and fully formed adducts. Confirmatory evidence is obtained from the fact that the entropy of activation of Diels–Alder retrogressions is small (see section 4,3).

It is not suggested that the ϱ^* factors of all Diels–Alder associations are larger than unity. In exceptional cases, for instance, in the reactions between maleic anhydride with isobutyl- or neopentylbutadiene, steric requirements may be responsible for a "freezing-in" of certain torsional modes of low frequency, in passing from initial to transition state, thereby reducing the numerical value of ϱ^*. It is possible that an effect of this kind contributes in making the kinetic A factors ref. nos. 30 and 31 in Table 14 abnormally small.

5.3 Two-step mechanism of uncatalysed thermal Diels–Alder reactions

It has been frequently suggested that thermal non-catalysed Diels–Alder reactions occur in two steps namely

$$a + b \rightleftharpoons x \rightleftharpoons e \tag{82}$$

where the intermediate x is a diradical. Only two arguments[175, 200] in favour of this mechanism are briefly discussed below, but in the first instance it is shown that retention of geometrical configuration of a dienophile and preferential formation of endo-adducts in certain Diels–Alder reactions (see Chap. 2) is consistent not only with the one step mechanism, but also with scheme (82), provided the formation of intermediate x is rate determining and the second step is sufficiently fast. If it is assumed that a Diels–Alder reaction with a *cis*- and *trans*-dienophile occurs according to scheme (82), intermediates* represented by (VI) and (VII) would be formed. There will be retention of the configuration of the dienophile, if bond formation between carbon atoms 4 and 6 is faster than rotation about the bond between carbon atoms 5 and 6. As a second example two intermediates, (VIII) and (IX) are shown which would both play a role if the dimerisation of cyclopentadiene

---

* For the sake of simplicity only one electronic configuration of species (VI)–(X) is shown. For a better representation of such intermediates see *e.g.* ref. 175 or 229. The two arrows in formulae (VI)–(X) indicated unpaired electrons; "intermediates" in which these two electrons are paired do not fall within the scope of a discussion of a two step mechanism.

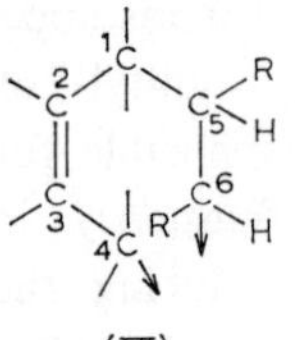

(VI)

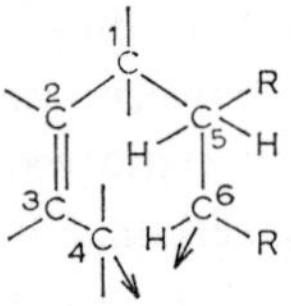

(VII)

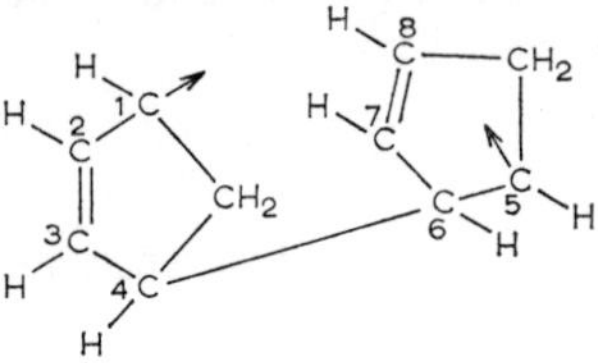

(VIII)

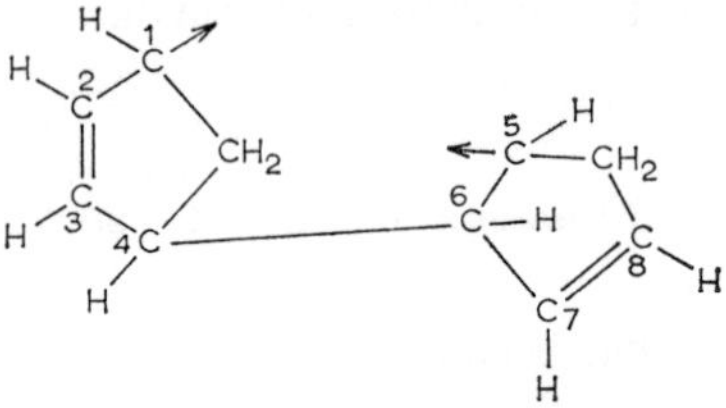

(IX)

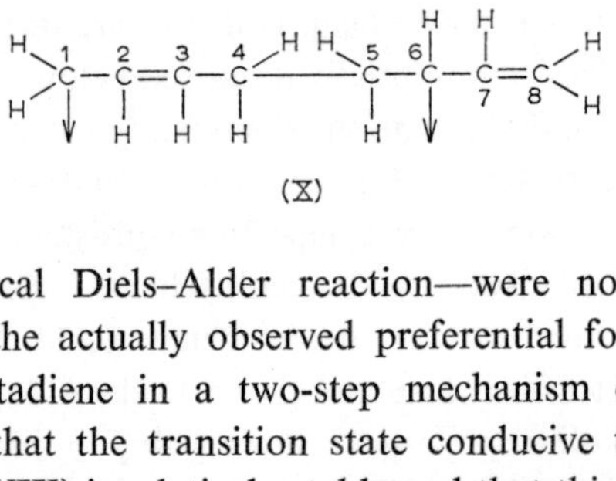

(X)

—a typical Diels–Alder reaction—were not stereospecific. To explain the actually observed preferential formation of endo-dicyclopentadiene in a two-step mechanism one would have to assume that the transition state conducive to the formation of species (VIII) is relatively stable and that this is brought about by non-bonding attraction essentially similar to that which has been invoked in section 1 to explain the different stabilities of the transition states shown in Figs. 3 and 4.

As stereochemical observations are not very useful to decide between the one and two-step mechanism other methods of approach have been employed. A careful investigation of the kinetics of dimerisation of gaseous butadiene has been carried out[175] and it has been suggested[175] that the rate determining step involves a transition state with only one incipient bond between the reactants and an open-chain diradical intermediate (X) which leads to vinylcyclohexene in a fast consecutive step in which bond formation between carbon atoms 1 and 6 occurs. A similar two-step mechanism has been proposed[200a] for two of the reactions in Table 16 in order to explain the influence of hydrostatic pressure on the rate. In support it has been claimed[153a,175,200] that the experimental entropy of activation, $\triangle S^*$, deduced from the kinetic A factor ref. no. 2, Table 13 and some of the experimental $\triangle V^*$ values of Table 16 indicate open chain transition states, rather than cyclic species. Revised calculations[229,199] show, however, that the observed $\triangle S^*$ value and the influence of hydrostatic pressure are also compatible with a one-step addition.

Table 20 shows that the kinetics of one Diels–Alder retrogression and of a number of diene associations are not influenced by various added substances known to react easily with free radicals. These results are evidence against the supposition that diradicals such as species (VI)–(X) play a role as intermediates. It could be suggested that the coupling between the unpaired electrons is sufficiently strong to produce triplet states and that therefore radical reagents are without effect. In this case, however, the kinetics of Diels–Alder associations a + b → c should be influenced by a restricted probability of singlet–triplet transition[230] and the principle of microscopic reversibility would require that the same effect operates in Diels–Alder retrogressions c → a + b. It has been observed that the kinetic A factors of Diels–Alder retrogressions are not abnormally low (see Table 19) and it can be concluded, therefore, that the A factors of the forward reactions a + b → c are also not substantially influenced by a restricted probability of electronic transitions. This is consistent with the statistical mechanical considerations in section 2. It is perhaps also of relevance that the added substance no. 8 in Table 20, which is known to accelerate singlet–triplet transitions, has no detectable influence on the rate of a diene association.

It appears, therefore, that the two-step mechanism of thermal uncatalysed Diels–Alder reactions is unsatisfactory. In the photosensitized associations mentioned in section 4,6, on the other hand, triplet state intermediates may play an important role.

5. 4 Diels–Alder retrogressions and isomerisation of endo–exo-adducts

It has been suggested[148] that the transition state in the non-dissociative rearrangements (22) and (23) on p. 36 is the same as that of the hypothetical Diels–Alder retrogressions into cyclopentadiene and hydroxycyclopentadiene. The postulated[148] intermediate (XI) contains one fully formed bond between diene and dienophile, the dashed lines representing partial bonds and the crosses "secondary attractive forces", which are not to be confused with the dotted lines in the structures of Figs. 2–8. This hypothesis has been competently criticised[151] and it has been rightly pointed

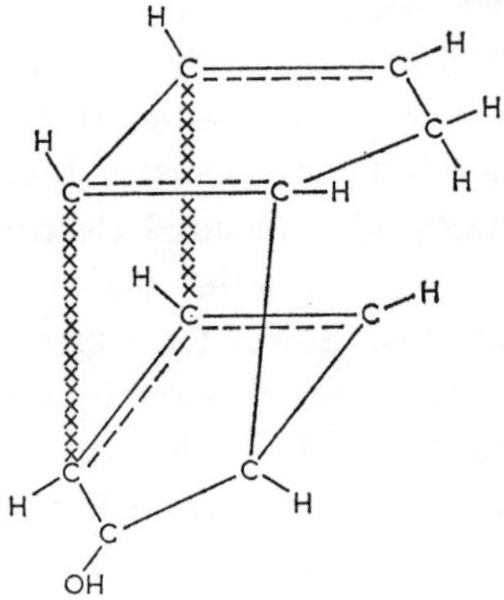

(XI)

out[231] that the reaction path of a Cope rearrangement and of a Diels–Alder retrogression need not be identical. Kinetic measurements to prove the non-identity have been carried out[157], but for reasons mentioned in section 2,4 the results are not yet conclusive. If, on the other hand, the identity of the reaction paths of a Cope rearrangement and of a Diels–Alder retrogression is to be made plausible, extensive kinetic tests under different conditions would be required. Experiments of this kind are still outstanding.

5. 5 Catalysed associations

It has been shown in Chap. 4 that the acid and phenol catalysis of a Diels–Alder reaction can be interpreted by initiation, propagation and termination steps, represented by (66)–(68), on p. 69 and by (74)–(76) on p. 70. In the halogeno-acetic or hydrochloric acid catalysed formation of dicyclopentadiene the proton transfer step, producing species HS$^+$...B$^-$ in the initiation (66), is rate-determining; this intermediate can be formulated thus[232]:

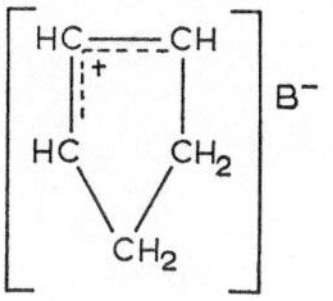

$$\left[\begin{array}{c} \text{structure XII} \end{array} \right] B^-$$

(XII)

where B$^-$ is the acid anion. In the propagation step (67) species (XII) combines with cyclopentadiene, thereby forming a bicyclopentyl derivative the structure of which is represented[232] by (XIII). This reacts in the termination step (68) to give dicyclopentadiene by elimination of acid, HB, and formation of a bond between carbons indicated by asterisks. The intermediate (XIII) can exist in the form of isomers, but in the catalysed Diels–Alder reactions (62) and (63) the only endo-adduct could be isolated* so far[212]. It can be tentatively assumed, therefore, that the isomer conducive to conversion into endo-dicyclopentadiene is preferentially formed (*cf.* section 5,3). Confirmatory evidence for the formation of an intermediate with one bond between the five carbon rings is the preparative isolation of a dicyclopentyl derivative[216]. It is probable that the transition state in the step leading to intermediate (XIII)

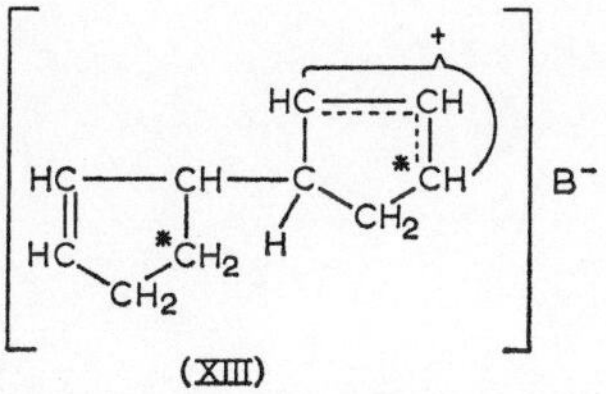

$$\left[\begin{array}{c} \text{structure XIII} \end{array} \right] B^-$$

(XIII)

---

* It would be worth while to test by vapour phase chromatography whether traces of exo-adducts are also formed.

contains one incipient bond between diene and dienophile and such transition states may also play a role in other reactions listed in Table 21. Effects which may be invoked to explain the stabilization of such transition states and the relatively small activation energies, E_c, in Table 21 have been mentioned in section 4,5.

APPENDIX

In Chapters 3, 4 and 5, entropy changes $\triangle S$ and $\triangle S^*$ are mentioned and it has been pointed out that low vibrational modes in Diels–Alder adducts or transition states enhance their statistical probability. It is useful, therefore, to give in Table 23, a list of standard formulae[158a] for entropy calculations.

TABLE 23
STANDARD FORMULAE FOR ENTROPY CALCULATIONS

Freedom	Contribution to S/R and S^*/R
Translation (relating to g, mol and cm³)	$1.299 + \ln \left\{ \left(\dfrac{T}{100°} \right)^{3/2} M^{3/2} \right\}$
External rotation	$3.46 + \ln \left\{ \dfrac{(I'I''I''')^{1/2}}{(10^{-40} \text{g cm}^2)^{3/2}} \left(\dfrac{T}{100°} \right)^{3/2} \dfrac{1}{\sigma} \right\}$
Harmonic vibration	$\dfrac{\Theta/T}{e^{\Theta/T}-1} + \ln \{(1-e^{-\Theta/T})^{-1}\}$
Internal rotation	$1.54 + \ln \left\{ \left(\dfrac{i}{10^{-40}\text{g cm}^2} \right)^{1/2} \left(\dfrac{T}{100°} \right)^{1/2} \dfrac{1}{\sigma_1} \right\}$

The symbols have the following meaning: M is the molecular weight on the conventional chemical scale, I', I'' and I''' are the principal moments of inertia, σ is the external symmetry number, Θ is the frequency of a harmonic internal vibration, expressed in terms of temperature, using the relation $h\nu = kT$, i is the reduced moment of inertia, and σ_1 is the symmetry number appropriate to the internal rotation. The principle moments of inertia I', I'' and I''' are defined as follows. If I is the moment of inertia of an atom or group of atoms of mass m with respect to a system of rectangular co-ordinates, x, y, and z, with arbitrary direction and with the centre of gravity of the molecule as origin, I', I'' and I''' are the roots of the secular equation given below:

$$\begin{vmatrix} \Sigma m(y^2+z^2) - I & \Sigma mxy & \Sigma mxz \\ \Sigma myz & \Sigma m(z^2+x^2) - I & \Sigma myz \\ \Sigma mxz & \Sigma myz & \Sigma m(x^2+y^2) - I \end{vmatrix} = 0$$

The computation of the vibrational contributions ϱ' and ϱ^* (see Table 10 and (81)) involves an *a priori* calculation of the expressions listed in the first, second and last lines of Table 23.

REFERENCES

1 K. ALDER, *Handbuch Biologische Arbeits-methoden*, ED. E. ABDERHALDEN, Urban and Schwarzenberg, Berlin, 1933, Abt. I, Teil II, 2 Hälfte, Band II; M. R. DELABY, *Bull. Soc. Chim. France* [5], 4 (1937) 765; S. LAMDAN. *Rev. Centro. Estud. Farm. bioquim.*, 29 (1939) 70; C. ELLIS, *The Chemistry of Synthetic Resins*, Reinhold, New York, 1935, Chapter 40; M. C. KLOETZEL, *Org. Reactions*, 4 (1948) 1; H. L. HOLMES, *Org. Reactions*, 4 (1948) 60; BUTZ and RYTINA, *Org. Reactions*, 5 (1948) 136; K. ALDER, *New Methods of Preparative Organic Chemistry*, New York, 1948, p. 381; S. S. NOVIKOV, G. A. SHVEKHGEIMER and A. A. DUDINSKAYA, *Usp. Khim.*, 29 (1960) 79; J. G. MARTIN and R. K. HILL, *Chem. Rev.*, 61 (1961) 537; J. K. STILLE, *Fortschr. Hochpolym. Forsch.*, 3 (1961) 48.
2 S.B. NEEDLEMAN and M.C. CHANG KUO, *Chem. Rev.*, 62 (1962) 405.
3 C. K. INGOLD, *Structure and Mechanism in Organic Chemistry*, Bell, London, 1953, Chapter XII.
4 C. K. INGOLD, The Transition State, *Chem. Soc. Special Publ.*, 1962, No. 16, page 119.
5 O. DIELS and K. ALDER, *Ann.*, 460 (1928) 98.
6 O. DIELS and K. ALDER, *Ber.*, 62 (1929) 2087.
7 E. H. FARMER and F. L. WARREN, *J. Chem. Soc.*, (1929) 897.
8 K. ALDER and G. STEIN, *Ann.*, 501 (1933) 247.
9 W. J. MIDDLETON, R. E. HECKERT, E. L. LITTLE Jr. and G. G. KRESPAN, *J. Am. Chem. Soc.*, 80 (1958) 2783.
10 M. OZOLINS and G. H. SCHENK, *Anal. Chem.*, 33 (1961) 1035.
11 K. ALDER and H. F. RICKERT, *Ann.*, 543 (1940) 1.
12 K. ALDER and E. WINDEMUTH, *Ber.*, 71 (1938) 1939.
13 K. ALDER and K. H. BACKENDORF, *Ber.*, 71 (1938) 2199.
14 A. A. PETROV, *Zh. Obshch. Khim.*, 11 (1941) 309.
15 D. V. NIGHTINGALE, M. MAIENTHAL and J. A. GALLAGHER, *J. Am. Chem. Soc.*, 75 (1953) 4852.
16 O. DIELS and K. ALDER, *Ber.*, 62 (1929) 2337.
17 G. O. SCHENCK, *Ber.*, 77 (1944) 741.
18 P. BRIGL and R. HERRMANN, *Ber.*, 71 (1938) 2280.
19 K. ALDER and H. F. RICKERT, *Ber.*, 72 (1939) 1983.
20 O. DIELS and K. ALDER, *Ann.*, 470 (1929) 62.
21 O. GRUMMITT and F. J. CHRISTOPH, *J. Am. Chem. Soc.*, 73 (1951) 3479.
22 C. F. H. ALLEN, C. G. ELIOT and A. BELL, *Can. J. Res.*, 17B (1939) 75.

REFERENCES

23 H. J. BACKER, *Rec. Trav. Chim.*, 58 (1939) 643.
24 H. J. BACKER and J. STRATING, *Rec. Trav. Chim.*, 56 (1937) 1069.
25 C. A. STEWART, *J. Am. Chem. Soc.*, 84 (1962) 117.
26 E. H. FARMER and F. L. WARREN, *J. Chem. Soc.*, (1931) 3221.
27 W. J. BAILEY and W. A. KLEIN, *J. Am. Chem. Soc.*, 79 (1957) 3124.
28 D. CRAIG, *J. Am. Chem. Soc.*, 65 (1943) 1006.
29 O. GRUMMITT and F. J. CHRISTOPH, *J. Am. Chem. Soc.*, 73 (1951) 3479.
30 W. H. CAROTHERS, I. WILLIAMS, A. M. COLLINS and J. E. KIRBY, *J. Am. Chem. Soc.*, 53 (1931) 4203.
31 H. P. KAUFMAN and J. BALTES, *Fette u. Seifen*, 43 (1936) 93.
32 J. C. ECK and E. W. HOLLINGSWORTH, *J. Am. Chem. Soc.*, 64 (1942) 140.
33 B. J. F. HUDSON and R. ROBINSON, *J. Chem. Soc.*, (1941) 715.
34 K. ALDER, F. PASCHER and H. VAGT, *Ber.*, 75 (1942) 1501.
35 T. WAGNER-JAUREGG, *Ann.*, 491 (1931) 1.
36 E. CLAR, *Ber.*, 65 (1932) 846.
37 R. SCHOLL and K. MEYER, *Ber.*, 67 (1934) 1236; *cf.* M. S. NEWMAN, *J. Am. Chem. Soc.*, 62 (1940) 1683.
38 K. ALDER and E. WINDEMUTH, *Ber.*, 71 (1938) 1939.
39 L. M. JOSHEL and L. W. BUTZ, *J. Am. Chem. Soc.*, 63 (1941) 3350; W. NUDENBERG and L. W. BUTZ, *J. Am. Chem. Soc.*, 66 (1944) 307.
40 H. STAUDINGER, *Ann.*, 467 (1928) 73; H. STAUDINGER and H. A. BRUSON, *Ann.*, 447 (1926) 97.
41 K. ALDER and G. STEIN, *Ann.*, 485 (1931) 223; 496 (1932) 204; *Ber.*, 67 (1934) 613.
42 E. A. PRILL, *J. Am. Chem. Soc.*, 69 (1947) 62.
43 K. ALDER and G. STEIN, *Angew. Chem.*, 50 (1937) 510.
44 W. DILTHEY, W. SCHOMMER and O. TRÖSKEN, *Ber.*, 66 (1933) 1627; W. DILTHEY, I. THEWALT and O. TRÖSKEN, *Ber.*, 67 (1934) 1959.
45 W. DILTHEY and G. HURTIG, *Ber.*, 67 (1934) 2004.
46 O. DIELS, K. ALDER and E. NAUJOKS, *Ber.*, 62 (1929) 554.
47 W. NUDENBERG and L. W. BUTZ, *J. Am. Chem. Soc.*, 66 (1944) 307.
48 L. MANDELL and W. A. BLANCHARD, *J. Am. Chem. Soc.*, 79 (1957) 2343, 6198.
49 C. F. H. ALLEN and J. W. GATES, *J. Am. Chem. Soc.*, 65 (1943) 1283.
50 D. B. CLAPP, *J. Am. Chem. Soc.*, 61 (1939) 2733.
51 J. L. MELLES, *Rec. Trav. Chim.*, 71 (1952) 869.
52 O. DIELS, W. KOCH and H. FROST, *Ber.*, 71 (1938) 1163.
53 W. J. BAILEY and W. B. LAWSON, *J. Am. Chem. Soc.*, 79 (1957) 1444.
54 A. WINDAUS and A. LÜTTRINGHAUS, *Ber.*, 64 (1931) 850.
55 K. ALDER and H. F. RICKERT, *Ber.*, 70 (1937) 1364.
56 O. DIELS, K. ALDER, and K. MÜLLER, *Ann.*, 490 (1931) 257.
57 E. P. KOHLER, M. TISHLER, H. POTTER and H. T. THOMPSON, *J. Am. Chem. Soc.* 61 (1939) 1057.
58 A. C. COPE and C. L. BUMGARDNER, *J. Am. Chem. Soc.*, 78 (1956) 2812.
59 A. T. BLOMQUIST and A. GOLDSTEIN, *J. Am. Chem. Soc.*, 77 (1955) 998.
60 R. C. COOKSON and N. S. WARIYAR, *J. Chem. Soc.*, (1957) 327.
61 W. H. JONES, D. MANGOLD and H. PLIENINGER, *Tetrahedron*, 18 (1962) 267.

REFERENCES

62 P. YATES and P. EATON, *J. Am. Chem. Soc.*, 82 (1960) 4436.
62a M. C. KLOETZEL, R. P. DAYTON and H. L. HERZOG, *J. Am. Chem. Soc.*, 72 (1950) 273.
63 D. BRYCE-SMITH and B. VICKERY, *Chem. and Ind.*, (1961) 429.
64 E. CLAR, *Ber.*, 65 (1932) 503.
65 W. E. BACHMANN and M. C. KLOETZEL, *J. Am. Chem. Soc.*, 60 (1938) 481.
66 K. ALDER and O. ACKERMANN, *Chem. Ber.*, 87 (1954) 1567.
67 A. T. BLOMQUIST and J. A. VERDOL, *J. Am. Chem. Soc.*, 77 (1955) 1806; 78 (1956) 109.
68 O. B. WEBSTER and W. H. SHARKEY, *J. Org. Chem.*, 27 (1962) 3354.
69 A. T. BLOMQUIST and Y. C. MEINWALD, *J. Am. Chem. Soc.*, 82 (1960) 3619.
70 A. T. BLOMQUIST and Y. C. MEINWALD, *J. Am. Chem. Soc.*, 81 (1959) 667.
71 A. T. BLOMQUIST, J. WOLINSKY, Y. C. MEINWALD, and D. LONGONE, *J. Am. Chem. Soc.*, 78 (1956) 6057.
72 W. J. BAILEY and H. R. GOLDEN, *J. Am. Chem. Soc.*, 75 (1953) 4780.
73 W. J. BAILEY and M. MADOFF, *J. Am. Chem. Soc.*, 75 (1953) 5603.
74 J. P. ECONOMY, *Dissertation Abstr.*, 16 (1956) 236.
75 H. HOPFF and A. K. WICK, *Helv. Chim. Acta*, 44 (1961) 19, 380.
76 M. P. CAVA and A. A. DEANA, *J. Am. Chem. Soc.*, 81 (1959) 4266.
77 F. R. JENSEN, W. E. COLEMAN and A. J. BERLIN, *Tetrahedron Letters*, (1962) 15.
78 R. KUHN and T. WAGNER-JAUREGG, *Ber.*, 63 (1930) 2662.
79 K. ALDER and H. V. BRACHEL, *Ann.*, 608 (1957) 195.
80 K. ALDER and M. SCHUMACHER, *Ann.*, 570 (1950) 178.
81 H. P. KAUFMANN and R. K. SUD, *Chem. Ber.*, 92 (1959) 2797.
82 T. W. CAMPBELL and R. N. MCDONALD, *J. Org. Chem.*, 24 (1959) 730.
83 W. J. BAILEY and N. A. NIELSEN, *J. Org. Chem.*, 27 (1962) 3088.
84 E. R. H. JONES, H. H. LEE and M. C. WHITING, *J. Chem. Soc.*, (1960) 341.
85 T. NOZOE, T. MUKAI, K. TAKASE and T. NAGASE, *Proc. Japan Acad.*, 28 (1952) 477.
86 K. ALDER and F. G. JACOBS, *Chem. Ber.*, 86 (1953) 1528; K. ALDER, K. KAISER and M. SCHUMACHER, *Ann.*, 602 (1957) 80.
87 M. AVRAM, G. MAKTEESCU and C. D. NENITZESCU, *Ann.*, 636 (1960) 174.
88 K. B. WIBERG and W. J. BARTLEY, *J. Am. Chem. Soc.*, 82 (1960) 6375.
89 M. A. BATTISTE, *Chem. and Ind.*, (1961) 550.
90 G. WITTIG, *Angew. Chem. Intern. Edn. Engl.*, 1 (1962) 415.
91 G. WITTIG and A. KREBS, *Ber.*, 94 (1961) 3260; G. WITTIG and R. POHLKE, *Ber.*, 94 (1961) 3276.
92 J. F. BUNNETT, *J. Chem. Educ.*, 38 (1961) 278.
93 H. HEANEY, *Chem. Rev.*, 62 (1962) 81.
94 R. HUISGEN, in H. H. ZEISS, *Organometallic Chemistry*, Reinhold, New York, 1960, p. 36.
95 G. WITTIG and L. POHMER, *Chem. Ber.*, 89 (1956) 1334.
96 G. WITTIG, *Org. Syn.*, 39 (1959) 75.
97 G. WITTIG, W. STILTZ and E. KNAUSS, *Angew. Chem.*, 70 (1958) 166.
98 G. WITTIG and R. W. HOFFMANN, *Angew. Chem.*, 73 (1961) 435.
99 G. WITTIG and H. F. EBEL, *Angew. Chem.*, 72 (1960) 564; *Ann.*, 650 (1961) 20.

100 G. Wittig and V. Wahl, *Angew. Chem.*, 73 (1961) 492.
101 T. Kauffmann and F. P. Boettcher, *Angew. Chem.*, 73 (1961) 65.
102 J. G. Aston, G. J. Szasz, H. W. Wooley and F. G. Brickwedde, *J. Chem. Phys.*, 14 (1946) 67; J. G. Aston, *Discussions Faraday Soc.*, 10 (1951) 73; G. J. Szasz and N. Sheppard, *Trans. Faraday Soc.*, 49 (1953) 358.
103 K. Alder and G. Stein, *Angew. Chem.*, 47 (1934) 837.
104 W. E. Bachmann and L. B. Scott, *J. Am. Chem. Soc.*, 70 (1948) 1458.
105 J. Sauer, H. Wiest and M. Mielert, *Z. Naturforsch.*, 17 (B) (1962) 203; J. Sauer, D. Lang and H. Wiest, *Z. Naturforsch.*, 17 (B) (1962) 206.
106 K. Alder and G. Stein, *Ann.*, 514 (1934) 1.
107 K. Alder and G. Stein, 501 (1933) 247; 504 (1933) 205, 216; see also R. C. Cookson, E. Crundwell and J. Hudec, *Chem. and Ind. (London)*, (1958) 1003.
108 H. Stobbe and F. Reuss, *Ann.*, 391 (1912) 151; E. G. V. Barrett and L. J. Burrage, *J. Phys. Chem.*, 37 (1933) 1029; K. W. F. Kohlrausch and R. Seka, *Ber.*, 69 (1936) 729; G. A. Benford and A. Wassermann, *J. Chem. Soc.*, (1939) 362.
109 A. Wassermann, *J. Chem. Soc.*, (1935) 828.
110 A. Wassermann, *J. Chem. Soc.*, (1935) 1511.
111 O. Diels and K. Alder, *Ann.*, 478 (1930) 137.
112 E. P. Kohler, M. Tishler, H. Potter and H. T. Thompson, *J. Am. Chem. Soc.*, 61 (1939) 1057; K. Alder and H. H. Mölls, *Chem. Ber.*, 89 (1956) 1960.
113 W. Reppe, O. Schlichting, K. Klager and T. Toepel, *Ann.*, 560 (1948) 1; M. Avram, G. Mateescu and C. D. Nenitzescu, *Ann.*, 636 (1960) 174.
114 K. Alder and R. Rührmann, *Ann.*, 566 (1950) 1.
115 K. Alder, F. W. Chambers and W. Trimborn, *Ann.*, 566 (1950) 27.
116 R. B. Woodward and H. Baer, *J. Am. Chem. Soc.*, 70 (1948) 1161.
117 J. A. Berson and R. Swidler, *J. Am. Chem. Soc.*, 75 (1953) 1721.
118 K. Alder and G. Stein, *Ann.*, 514 (1934) 197; J. D. Roberts, E. R. Trumbull Jr., W. Bennett and R. Armstrong, *J. Am. Chem. Soc.*, 72 (1950) 3122.
119 K. Alder, K. Heimbach and R. Reubke, *Chem. Ber.*, 91 (1958) 1516.
120 K. Alder and H. F. Rickert, *Ann.*, 543 (1950) 1.
121 J. A. Berson, Z. Hamlet and W. A. Mueller, *J. Am. Chem. Soc.*, 84 (1962) 297.
122 K. Alder and H. Vagt, *Ann.*, 571 (1951) 153.
123 R. L. Frank, R. D. Emmick and R. S. Johnson, *J. Am. Chem. Soc.*, 69 (1947) 2313; D. Craig, *J. Am. Chem. Soc.*, 72 (1950) 1678.
124 K. Alder and M. Schumacher, *Ann.*, 571 (1951) 87.
125 K. Alder, M. Schumacher and O. Wolff, *Ann.*, 564 (1949) 79.
126 K. Alder and W. Vogt, *Ann.*, 564 (1949) 120.
127 J. S. Meek, B. T. Poon, R. T. Merrow and S. J. Cristol, *J. Am. Chem. Soc.*, 74 (1952) 2669.
128 R. B. Woodward and H. Baer, *J. Am. Chem. Soc.*, 66 (1944) 645; D. Craig, *J. Am. Chem. Soc.*, 73 (1951) 4889.
129 K. Alder and G. Stein, *Ber.*, 67 (1934) 613.

130 E. W. J. BUTZ and L. W. BUTZ, *J. Org. Chem.*, 7 (1942) 199; M. MOUSSERON, F. WINTERNITZ and G. ROUZIER, *Compt. Rend.*, 238 (1954) 1661.
131 M. C. KLOETZEL and H. L. HERZOG, *J. Am. Chem. Soc.*, 72 (1950) 1991; K. TAKEDA, S. NAGAKURA and K. KITAHONOKI, *Pharm. Bull. (Tokyo)*, 1 (1953) 135.
132 E. LEHMANN and W. PAASCHE, *Ber.*, 68 (1935) 1068; K. ALDER and W. VOGT, *Ann.*, 564 (1949) 109, 120.
133 K. ALDER and M. SCHUMACHER, *Ann.*, 565 (1949) 148.
134 H. FIESSELMANN, *Ber.*, 75 (1942) 881.
135 S. HÜNIG and H. KAHANEK, *Ber.*, 90 (1957) 238.
136 J. S. MEEK and B. T. TRAPP, *J. Am. Chem. Soc.*, 74 (1952) 2686.
137 K. ALDER, H. VAGT and W. VOGT, *Ann.*, 565 (1949) 135; K. ALDER and J. HAYDN, *Ann.*, 570 (1950) 201; J. S. MEEK, D. E. RAMEY and S. J. CRISTOL, *J. Am. Chem. Soc.*, 73 (1951) 5563.
138 C. S. MARVEL and N. O. BRACE, *J. Am. Chem. Soc.*, 71 (1949) 37; H. R. SNYDER and G. I. POOS, *J. Am. Chem. Soc.*, 72 (1950) 4104.
139 K. ALDER, M. SCHUHMACHER and O. WOLFF, *Ann.*, 570 (1950) 230; E. A. BRAUDE and E. A. EVANS, *J. Chem. Soc.*, (1956) 3238; C. C. J. CULVENOR and T. A. GEISSMANN, *Chem. and Ind.*, (1959) 366.
139a I. N. NAZAROW, Y. A. TITOV and A. I. KUZNETSOVA, *Dokl. Akad. Nauk SSSR.*, 124 (1959) 586; *Chem. Abstr.*, 53 (1959) 11268; M. E. HENNIS, *J. Org. Chem.*, 28 (1963) 2370.
140 K. ALDER, G. STEIN, J. REESE and W. GRASSMANN, *Ann.*, 496 (1932) 204; K. ALDER and E. WINDEMUTH, *Ber.*, 71 (1938) 2409; K. ALDER and H. F. RICKERT, *Ann.*, 543 (1939) 1.
141 R. E. LIDOV, H. BLUESTONE and S. B. SOLOWAY, *Agricultural Controlled Chemicals; Advances Chemistry Series Am. Chem. Soc.*, 1 (1950) 175.
142 S. B. SOLOWAY, *J. Am. Chem. Soc.*, 74 (1952) 1027.
143 A. WASSERMANN, *J. Chem. Soc.*, (1936) 1028.
144 K. ALDER, F. H. FLOCK and P. JANSSEN, *Chem. Ber.*, 89 (1956) 2689; O. DIELS, K. ALDER and G. STEIN, *Ber.*, 62 (1929) 2337; *Ann.*, 501 (1933) 247.
145 R. K. HILL and J. G. MARTIN, *Proc. Chem. Soc.*, (1959) 390.
146 L. DE VRIES, R. HECK, R. PICCOLINI and S. WINSTEIN, *Chem. and Ind.*, (1959) 1416.
147 C. E. H. BAWN, *Proc. Chem. Soc.*, (1962) 165.
148 R. B. WOODWARD and T. J. KATZ, *Tetrahedron*, 6 (1959) 70.
148a R. C. COOKSON, J. HUDEC and R. O. WILLIAMS, *Tetrahedron Letters*, 22 (1960) 29; P. YATES and P. EATON, *Tetrahedron*, 12 (1961) 13.
149 J. A. BERSON and D. A. BEN-EFRAIM, *J. Am. Chem. Soc.*, 81 (1959) 4083.
150 J. A. BERSON, J. S. WALIA, A. REMANICK, S. SUZUKI, P. REYNOLDS-WARNHOFF and D. WILLNER, *J. Am. Chem. Soc.*, 83 (1961) 3986.
151 J. A. BERSON and A. REMANICK, *J. Am. Chem. Soc.*, 83 (1961) 4947.
152 J. A. BERSON and W. A. MUELLER, *J. Am. Chem. Soc.*, 83 (1961) 4940.
153 R. P. LUTZ and J. D. ROBERTS, *J. Am. Chem. Soc.*, 83 (1961) 2198.
153a C. WALLING and H. C. SCHUGER, *J. Am. Chem. Soc.*, 85 (1963) 601.
154 See, *e.g.* A. C. COPE, C. M. HOFMANN and E. M. HARDY, *J. Am. Chem. Soc.*, 63 (1941) 1852.

REFERENCES

155 E. G. FOSTER, A. C. COPE and F. DANIELS, *J. Am. Chem. Soc.*, 69 (1947) 1893.
155a C. WALLING and M. NAIMAN, *J. Am. Chem. Soc.*, 84 (1962) 2628.
156 J. A. BERSON, R. D. REYNOLDS and W. M. JONES, *J. Am. Chem. Soc.*, 78 (1956) 6049.
157 J. E. BALDWIN and J. D. ROBERTS, *J. Am. Chem. Soc.*, 85 (1963) 115.
158 J. H. VAN 'T HOFF, *Lectures on Theoretical and Physical Chemistry*, 1898, Vol. I, p. 231; O. DIMROTH, *Ann.*, 377 (1910) 134.
158a See, for instance, R. H. FOWLER and E. H. GUGGENHEIM, *Statistical Thermodynamics*, Cambridge, 1939.
159 G. A. BENFORD and A. WASSERMANN, *J. Chem. Soc.*, (1939) 367.
160 G. W. WHELAND, *Resonance in Organic Chemistry*, Wiley, New York, 1955, Table 3, 7.
161 R. D. BROWN, *Quart. Rev.*, 6 (1952) 63; *J. Chem. Soc.*, (1950) 691, 2730; (1951) 1612.
162 A. WASSERMANN, *Proc. Roy. Soc. (London)*, A 178 (1941) 370.
163 D. ROWLEY and H. STEINER, *Discussions Faraday Soc.*, 10 (1951) 198.
164 H. W. B. REED, *J. Chem. Soc.*, (1951) 685.
165 D. D. COFFMAN, P. L. BARRICK, R. D. CRAMER and M. S. RAASCH, *J. Am. Chem. Soc.*, 71 (1949) 490.
166 J. J. DRYSDALE, W. W. GILBERT, H. K. SINCLAIR and W. H. SHARKEY, *J. Am. Chem. Soc.*, 80 (1958) 245, 3672. See also J. D. ROBERTS and J. D. SHARTS in *Organic Reactions*, Vol. 12, Wiley, New York, 1962, p. 1.
167 C. A. STEWART, *J. Am. Chem. Soc.*, 84 (1962) 117.
168 A. WASSERMANN, *Stereochemie*, Ed. K. FREUDENBERG, Deuticke, Vienna, 1933, p. 781.
168a P. D. BARTLETT, L. K. MONTGOMERY and B. SEIDEL, *J. Am. Chem. Soc.*, 86 (1964) 616; L. K. MONTGOMERY, K. SCHUELLER and P. D. BARTLETT, *J. Am. Chem. Soc.*, 86 (1964) 626; P. D. BARTLETT and L. K. MONTGOMERY, *J. Am. Chem. Soc.*, 86 (1964) 628.
168b S. GLASSTONE, K. J. LAIDLER, and H. EYRING, *Theory of Rate Processes*, McGRAW-HILL, New York, 1941, p. 195–199.
169 R. N. PEASE, *J. Am. Chem. Soc.*, 51 (1929) 3470; P. SCHLÄPFER and M. BRUNNER, *Helv. Chim. Acta*, 13 (1930) 1125; H. A. TAYLOR and A. van HOOK, *J. Physical Chem.*, 39 (1939) 811.
170 M. W. TRAVERS, *Trans. Faraday Soc.*, 32 (1936) 236; 33 (1937) 1345.
171 R. N. PEASE, *J. Am. Chem. Soc.*, 52 (1930) 1158; 53 (1931) 613; H. H. STORCH, *J. Am. Chem. Soc.*, 56 (1934) 374; F. R. RUSSELL and H. C. HOTTEL, *Ind. Eng. Chem.*, 30 (1938) 183.
172 M. V. KRAUZE, M. S. NEMTSOV and E. A. SOSKINA, *Compt. Rend. Acad. Sci. U.R.S.S.*, 2 (1934) 305; 3 (1934) 262; *J. Gen. Chem. U.S.S.R.*, 5 (1935) 343, 356, 382; S. P. MITZENGENDLER, *J. Gen. Chem. U.S.S.R.*, 7 (1937) 1848.
173 W. E. VAUGHAN, *J. Am. Chem. Soc.*, 55 (1933) 4109.
174 J. B. HARKNESS, G. B. KISTIAKOWSKY and W. H. MEARS, *J. Chem. Physics*, 5 (1937) 682.
175 G. B. KISTIAKOWSKY and W. W. RANSOM, *J. Chem. Phys.*, 7 (1939) 725.
176 G. A. BENFORD and A. WASSERMANN, *J. Chem. Soc.*, (1939) 362.

REFERENCES

177 G. B. KISTIAKOWSKY and J. R. LACHER, *J. Am. Chem. Soc.*, 58 (1936) 123.
177a K. E. LEWIS and H. STEINER, *J. Chem. Soc.*, (1964) 3080.
178 B. S. KHAMBATA and A. WASSERMANN, *J. Chem. Soc.*, (1939) 371.
179 G. A. BENFORD, H. KAUFMANN, B. S. KHAMBATA and A. WASSERMANN, *J. Chem. Soc.*, (1939) 381.
180 H. KAUFMANN and A. WASSERMANN, *J. Chem. Soc.*, (1939) 870.
181 A. WASSERMANN, *J. Chem. Soc.*, (1935) 831.
182 A. WASSERMANN, *J. Chem. Soc.*, (1942) 623.
183 R. A. FAIRCLOUGH and G. N. HINSHELWOOD, *J. Chem. Soc.*, (1938) 236.
184 D. CRAIG, J. J. SHIPMAN and R. B. FOWLER, *J. Am. Chem. Soc.*, 83 (1961) 2885.
185 L. J. ANDREWS and R. M. KEEFER, *J. Am. Chem. Soc.*, 77 (1955) 6284.
186 J. SAUER, D. LANG and A. MIELERT, *Angew. Chem.*, 74 (1962) 352.
187 B. EISLER and A. WASSERMANN, *J. Chem. Soc.*, (1952) 978.
188 W. RUBIN and A. WASSERMANN, *J. Chem. Soc.*, (1950) 2205.
189 J. SAUER and H. WIEST, *Angew. Chem.*, 74 (1962) 353.
190 A. RODGMAN and G. F. WRIGHT, *J. Org. Chem.*, 18 (1953) 465.
191 G. N. HINSHELWOOD, *Trans. Farady Soc.*, 34 (1938) 105.
192 R. A. FAIRCLOUGH and G. N. HINSHELWOOD, *J. Chem. Soc.*, (1937) 538; (1938) 236.
193 M. G. EVANS and M. POLANYI, *Trans. Faraday Soc.*, 32 (1936) 1333; 33 (1937) 166; R. P. BELL, *Trans. Faraday Soc.*, 33 (1937) 496; J. A. V. BUTLER, *Trans. Faraday Soc.*, 33 (1937) 229; I. M. BARCLAY and J. A. V. BUTLER, *Trans. Faraday Soc.*, 34 (1938) 1445.
194 J. GILLOIS-DOUCET and P. RUMPF, *Bull. Soc. Chim. France*, (1959) 1823.
194a D. E. VAN SICKLE and J. O. RODIN, *J. Am. Chem. Soc.*, 86 (1964) 3091.
195 B. RAISTRICK, R. H. SHAPIRO and D. M. NEWITT, *J. Chem. Soc.*, (1939) 1761.
196 D. M. NEWITT and A. WASSERMANN, *J. Chem. Soc.*, (1940) 735.
197 E. RABINOWISCH, *Trans. Faraday Soc.*, 33 (1937) 1225; R. H. FOWLER and N. B. SLATER, *Trans. Faraday Soc.*, 34 (1938) 81.
198 M. G. EVANS and M. POLANYI, *Trans. Faraday Soc.*, 31 (1935) 875.
199 S. W. BENSON and J. A. BERSON, *J. Am. Chem. Soc.*, 84 (1962) 152.
200 C. WALLING and J. PEISACH, *J. Am. Chem. Soc.*, 80 (1958) 5819.
200a J. SAUER, D. LANG and H. WIEST, *Chem. Ber.*, 97 (1964) 3208.
201 L. KÜCHLER, *Nachr. Akad. Wissensch. Göttingen, Math. Physik. Kl. III*, 1 (1939) 231.
202 B. S. KHAMBATA and A. WASSERMANN, *J. Chem. Soc.*, (1939) 375.
203 S. SELZER, *J. Am. Chem. Soc.*, 85 (1963) 1360.
204 R. P. BELL, *The Proton in Chemistry*, Methuen, London, 1959, Chapter XI.
204a V. J. SHINER JR., *J. Am. Chem. Soc.*, 75 (1953) 2925; 76 (1954) 1603.
204b A. STREITWIESER, R. H. JAGOW, R. C. FAHEY, and S. SUZUKI, *J. Am. Chem. Soc.*, 80 (1958) 2326.
205 H. STOBBE and F. REUSS, *Ann.*, 391 (1912) 151; G. R. SCHULTZE, *J. Am. Chem. Soc.*, 56 (1934) 1552.
206 W. A. WATERS, *The Chemistry of Free Radicals*, Clarendon Press, Oxford, 1948, p. 43–46 and 176.

207 L. A. K. STAVELEY and C. N. HINSHELWOOD, *Proc. Roy. Soc. (London)*, 154 (1936) 335; *J. Chem. Soc.*, (1937) 1568; F. O. RICE and O. L. POLLY, *J. Chem. Physics*, 6 (1938) 273.
208 E. COLLINSON, F. S. DAINTON, D. R. SMITH, G. J. TRUDEL and S. TAZUKÉ, *Discussions Faraday Soc.*, 29 (1960) 188.
209 Unpublished work by G. S. HAMMOND, see ref. 152, footnote page 4941.
210 A. WASSERMANN, *J. Chem. Soc.*, (1946) 1089.
211 W. RUBIN, H. STEINER and A. WASSERMANN, *J. Chem. Soc.*, (1949) 3046.
212 A. WASSERMANN, *J. Chem. Soc.*, (1942) 618.
213 L. E. GAST, E. W. BELL and H. M. TEETER, *J. Am. Oil Chemists Soc.*, 33 (1956) 278.
214 E. H. INGOLD and A. WASSERMANN, *Trans. Faraday Soc.*, 35 (1939) 1022; B. S. KHAMBATA and A. WASSERMANN, *J. Chem. Soc.*, (1946) 1090.
214a E. P. LUTZ and G. H. BAILEY, *J. Am. Chem. Soc.*, 86 (1964) 3899.
215 R. ROBINSON and G. I. FRAY, British Patent 835, 840 (1960); P. YATES and P. EATON, *J. Am. Chem. Soc.*, 82 (1960) 4436; G. I. FRAY and R. ROBINSON, *J. Am. Chem. Soc.*, 83 (1961) 249; C. F. M. ALLEN, W. R. RYAN and J. A. VAN ALLAN, *J. Org. Chem.*, 27 (1962) 778.
216 C. F. BLAKELEY, R. J. GILLESPIE, L. ROUBINEK, A. WASSERMANN and R. F. M. WHITE, *J. Chem. Soc.*, (1961) 1939; C. F. BLAKELEY and A. WASSERMANN, *J. Chem. Soc.*, (1961) 1946.
217 G. S. HAMMOND, P. A. LEERMAKERS and N. J. TURRO, *J. Am. Chem. Soc.*, 83 (1961) 2397.
218 G. S. HAMMOND, N. J. TURRO and A. FISCHER, *J. Am. Chem. Soc.*, 83 (1961) 4674.
219 G. S. HAMMOND and R. S. H. LIU, *J. Am. Chem. Soc.*, 85 (1963) 477.
220 G. S. HAMMOND and W. M. HARDHAM, *Proc. Chem. Soc.*, (1963) 63.
221 W. G. HERKSTROETER, J. SALTIEL and G. S. HAMMOND, *J. Am. Chem. Soc.*, 85 (1963) 482.
222 D. BRYCE-SMITH and J. E. LODGE, *J. Chem. Soc.*, (1962) 2675; see also H. J. F. ANGUS and D. BRYCE-SMITH, *J. Chem. Soc.*, (1960) 4791; E. GROVENSTEIN, D. V. RAO and J. W. TAYLOR, *J. Am. Chem. Soc.*, 83 (1961) 1705.
223 M. G. EVANS and E. WARHURST, *Trans. Faraday Soc.*, 34 (1938) 614; M. G. EVANS, *Trans. Faraday Soc.*, 15 (1939) 824.
224 M. G. ETTLINGER and E. S. LEWIS, *Texas J. Sci.*, 14 (1962) 58.
225 A. WASSERMANN, *J. Chem. Soc.*, (1936) 432; *Trans. Faraday Soc.*, 35 (1939) 841.
226 C. K. INGOLD, ref. 3, p. 565 and various applications in Chapters 10 and 12.
226a G. S. HAMMOND, *J. Am. Chem. Soc.*, 77 (1955) 334.
227 L. HORNER and H. MERZ, *Ann.*, 570 (1950) 89; L. HORNER and W. SPIETSCHKA, *Ann.*, 579 (1953) 159; R. H. BURNELL and W. I. TAYLOR, *J. Chem. Soc.*, (1954) 3636; (1955) 2054.
228 C. A. COULSON and P. L. DAVIES, *Trans. Faraday Soc.*, 48 (1952) 777.
228a P. DE SANTIS, E. GIGLIO, A. M. LIQUORI and A. RIPAMONTI, *J. Polymer Sci.*, 1 (1963, A) 1383.
229 A. WASSERMANN, *J. Chem. Soc.*, (1942) 612.

230 B. G. GOWENLOCK, *Quart. Rev.*, 14 (1960) 133; there also references to earlier work.
231 M. J. S. DEWAR, *Tetrahedron Letters*, No. 4, (1959) 16.
232 C. F. BLAKELEY and A. WASSERMANN, *J. Chem. Soc.*, (1961) 1946.

SUBJECT INDEX

Acetylene dicarboxylic acid, reaction with N-benzylpyrrole, 11
—, — with butadiene, 5
Acetylenic compounds as dienophiles, 5, 11, 12
Acid catalysis, 69, 96
Acids, $\alpha\beta$-unsaturated, reaction with butadiene, 3
Acrolein, adducts with butadienes, 30
—, reaction with cyclopentadiene, 56
—, — with 1-diethylaminobutadiene, 85
—, — with 3-methoxybutadiene, 85
Acrylates, adducts with butadienes, 30
Acrylonitrile, reaction with butadiene, 4
Activation energy, of one-step mechanism, 74
—, of retrogressions, 62
Activation entropy of retrogressions, 62, 92
Activation parameters, in catalysis, 70
—, of reactions in condensed state, 58
Acyclic dienes, reaction with maleic anhydride, 6
Added substances without influence, 64
Aldehydes, $\alpha\beta$-unsaturated, reaction with butadiene, 3
Aldrin, 9, 31
Alignment, mode of, 46
Allene, reaction with maleic anhydride, 15
Allyl compounds as dienophiles, 5
Anthracene, reaction with dienophiles, 14
—, — with maleic anhydride, 51, 91

Anthracene–maleic anhydride adducts, dissociation of, 14
Arenes, preparation of, 15
—, reaction with dienophiles, 14
Azodicarboxylates, dialkyl, reaction with cyclopentadiene, 51

Benzanthrene, reaction with dienophiles, 9
Benzene, reaction with maleic anhydride, 13
—, — with maleic anhydride triplets, 73
Benzocyclobutene, isomerisation of, 17
—, reaction with maleic anhydride, 17
Benzoquinone, as proton- and electron-acceptor, 70
—, reaction with butadiene, 3
—, — with cyclopentadiene, 54, 65, 79, 80, 86
—, — with 2,3-divinylbutadiene, 19
N-Benzylpyrrole, reaction with acetylene dicarboxylic acid, 11
Benzyne, 11, 21, 22, 23
1,1′-Bicyclohexenyl, reaction with dienophiles, 7
Bicyclo[3,2,2]nonanes, formation of, 13
Bicyclo[2,2,2]octadienes, formation of, 12
Biphenyl, 8
Butadiene, dimerisation of, 81, 94
—, reaction with cyclopropene, 20
—, — with dienophiles, 3, 4, 5
—, — with ethylene, 76, 84

SUBJECT INDEX

Butadiene triplets, 71
Butadienes, disubstituted, reactivity of, 6
—, monosubstituted, adducts with acrolein or acrylates, 30

Carbonyl compounds as photosensitizers, 71
Catalysed associations, 96
Catalysis of thermal reactions, 65
Cholestadienes, reaction with maleic anhydride, 7
Cisoid conformation, of dienes, 2, 89
—, sterically hindered formation of, 5, 6, 18
Cisoid dienes, reactivity of, 24
Cis-trans isomeric triplets of conjugated dienes, 72
Conformation of diene, 89
Conjugated dienes, cis-trans isomeric triplets of, 72
Cope rearrangements, 37, 96
Crotyl esters, isomerisation of, 37
Cumulenes, reaction with dienophiles, 19
Cycloalkenes as dienophiles, 20
Cycloalkynes as dienophiles, 20
Cyclobutane derivatives, formation of, 46
Cyclodeca-1,3-dienes, 13
Cycloheptadiene, reaction with dienophiles, 13
Cycloheptatriene, reaction with maleic anhydride, 19
Cyclohexadienes, formation of, 13
—, reaction with acetylenic dienophiles, 12
—, — with dienophiles, 12
—, — with maleic anhydride, 83
—, structure determination of, 12
Cyclohexane, formation of, 76
Cycloocta-1,3-diene, reaction with maleic anhydride, 13
Cyclooctatetraene, reaction with maleic anhydride, 19
Cyclooctyne, reaction with 2,5-diphenyl-3,4-benzofuran, 21

Cyclopentadiene, dimerisation of, 9, 45, 54, 56, 77, 78, 93
—, —, activation parameters of, 57
—, —, catalysed, 69, 97
—, geometrically isomeric adducts from, 58, 59
—, polymerisation of, 31
—, reaction with acrolein, 56
—, — with benzoquinone, 54, 65, 79, 80, 86
—, — with cyclopropene, 20
—, — with dialkyl azodicarboxylates, 51
—, — with dienophiles, 9, 13
—, — with dimethyl maleate or fumarate, 24
—, — with maleic anhydride, 3, 83
—, — with methyl methacrylate, 81, 82, 88
Cyclopentadienones, reaction with dienophiles, 10
Cyclopropene, reaction with butadiene, 20
—, — with cyclopentadiene, 20

Deuterium isotope effect, 55
1,2-Dibenzoylethylene, geometrically isomeric adducts from, 61
Dicyclopentadiene, dissolution of, 45
Dieldrin, 9, 31
Diene component, structure of, 1
1,3-Dienes, analysis of, 4
Dienophile, structure of, 2
Dienophile C=C bond, protection of, 15
Dienophiles, reactions with, see under the pertinent compounds
1-Diethylaminobutadiene, reaction with acrolein, 85
Dimerisation of cyclopentadiene, catalyzed, 69, 97
9,10-Dimethylanthracene, geometrically isomeric adducts from, 59
—, reaction with maleic anhydride, 51
2,3-Dimethylbutadiene, geometrically isomeric adducts from, 61
Dimethylenecycloalkanes, reaction with dienophiles, 15

SUBJECT INDEX

1,2-Dimethylenecyclobutane, reaction with maleic anhydride, 15
1,2-Dimethylenecyclohexa-3,5-diene, reaction with dienophiles, 17
1,2-Dimethylenecyclohexane, reaction with benzoquinone, 16
—, — with maleic anhydride, 16
1,2-Dimethylenecyclopentane, reaction with maleic anhydride, 16
1,2-Dimethylene-3,4-diphenylcyclobutane, reaction with maleic anhydride, 16
1,2-Dimethylene-3,4-diphenylcyclobut-3-ene, reaction with maleic anhydride, 16
Dimethyl fumarate, reaction with cyclopentadiene, 24
Dimethyl maleate, reaction with cyclopentadiene, 24
2,3-Dimethylnaphthalene, reaction with maleic anhydride, 14
2,5-Diphenyl-3,4-benzofuran, reaction with cyclooctyne, 21
1,1-Diphenylethylene, reaction with maleic anhydride, 8
Dipolar intermediate, 54
Dipole-dipole repulsion, 89
Dipole induction, 86
Distortion energy, 83
2,3-Divinylbutadiene, reaction with benzoquinone, 19
1,2-Divinylcyclobutane, formation of, 71

Electron transfer, 68, 70
Electronic energy of planar transition states, 91
Endo- and exo-isomeric adducts, formation of, 27, 28
Endo-dicyclopentadiene, formation of, 94, 97
Endo orientation of substituents, 2, 26
Entropy, calculations, standard formulae for, 99
—, changes, 42
—, of activation in one-step mechanism, 92

—, of reaction, 44
3,4-Epoxycyclohexene, formation of, 11
Equilibria in solution, 45
Equilibrium constant, in Diels-Alder reactions, 39
—, of cis- and transoid conformations, 24, 29
Esters, $\alpha\beta$-unsaturated, reaction with butadiene, 3
Ethylene, reaction with butadiene, 76, 84
—, — with furan, 11
Ethylenic compounds as dienophiles, 5
Exo orientation of substituents, 2, 26

Fire resistant paints, 9
o-Fluorobromobenzene, dehalogenation of, 22
Fulvenes, reaction with dienophiles, 10
Fumaronitrile, reaction with butadiene, 4
Furan, reaction with ethylene, 11
—, — with maleic anhydride, 11

Geometrically isomeric adducts, formation of, 58, 86, 89

Heat changes, 42
Heats of reaction, 43
— of solution, 43
Heptaphenylcycloheptatriene, formation of, 21
Hexachlorocyclopentadiene, reaction with maleic anhydride, 9, 89
Hexaethylidenecyclohexane, 17
Hexamethylradialene, 17
Hydrostatic pressure, 57

Indene, reaction with maleic anhydride, 8
Insecticides, 9, 31
Intermediates, structure identification of unstable, 15
Isobutylbutadiene, reaction with maleic anhydride, 93

SUBJECT INDEX

Isoeugenol methyl ether, reaction with maleic anhydride, 8
Isomerisation of adducts, 34
— of endo-exo adducts, 96
Isoprene triplets, 72
Isosafrole, reaction with maleic anhydride, 8, 84
Isotope effect in retrogressions, 62

Ketones, $\alpha\beta$-unsaturated, reaction with butadiene, 3
Kinetics, 48
— of catalysis, 66
Kinetic parameters, influence of solvents on, 54
—, of gaseous reactions, 50
—, of reactions in condensed state, 52
—, of retrogressions, 63

London dispersion effects, 86

Maleic anhydride, reactions with, see under the pertinent compounds
Maximum accumulation of unsaturation, 29
Methacrolein dimer, deuterium labelled, 37
3-Methoxybutadiene, reaction with acrolein, 85
α-Methylallyl group, formation from crotyl, 37
Methylarenes, formation of, 19
Methyl methacrylate, reaction with cyclopentadiene, 81, 82, 88
β-Methylstyrene, reaction with maleic anhydride, 84

Naphthacene, reaction with dienophiles, 14
Naphthalene, reaction with maleic anhydride, 14
Naphthalene-2-sulphonic acid as catalyst, 69
Neopentylbutadiene, reaction with maleic anhydride, 93
1-Nitronaphthalene, reaction with maleic anhydride, 14
Non-bonding attraction, 86, 94

Non-planar transition states, 90
Norbornene derivatives, 35
Norcamphane rings, adducts containing fused, 31, 32
Norcarane, formation of, 20

Octahydroanthraquinones, formation of, 3
Octahydrodimethanonaphthalene, formation of, 31
Octahydronaphthalenes, dehydrogenation of, 15
Octa-2,4,6-triene, reaction with maleic anhydride, 18
Olefins, preparation of, 15
One-step mechanism, 74, 92
Optically active adducts, 35

Paints, fire resistant, 9
β-Parinaric acid, reaction with maleic anhydride, 18
Partition functions, 40, 43, 92
Pentacene, formation of, 16
—, reaction with dienophiles, 14
1,3-Pentadiene, formation of triplet of, 71
Perylene, reaction with dienophiles, 9
α-Phellandrene, reaction with dienophiles, 12
Phenanthrene, 9
—, reaction with dienophiles, 14
Phenol catalysis, 70, 96
9-Phenylanthracene adduct, 35
1-Phenylbuta-1,3-diene, reaction with maleic anhydride, 5
Photo-associations, 71, 95
Polyenes, reaction with dienophiles, 18
Potential energy of reactants and product, 75
Proton transfer, 68, 69, 97
Pyridyne, 23
α-Pyrones, reaction with maleic anhydride, 13
Pyrroles, reaction with dienophiles, 11

Quinol, reaction with maleic anhydride, 13

SUBJECT INDEX

Quinol monomethyl ether, reaction with maleic anhydride, 13
p-Quinquephenyl, synthesis of, 18

Rates of reaction between dienes and cis-trans isomers, 25
Repulsion energy, 84
Resonance energies, 43, 84
Retrogressions, 61
—, of endo-exo adducts, 96
Ring strain, 43

Secondary attractive forces, 96
Stabilization of transition states, 85, 89
Stereochemical principles, 24ff
Stereochemistry of adducts, role of non-bonding attraction, 87
Stereospecific polymerisations, 34
Stereospecificity of reactions, 87
Steric hindrance, 5, 6, 18, 83
Steroids, identification of cisoid 1,3-diene system in, 12
Stilbene, reaction with maleic anhydride, 7
Styrene, reaction with maleic anhydride, 7, 84
—, — with tetrachloro-*o*-benzoquinone, 85

Temperature range, 3
α-Terpinene, reaction with dienophiles, 12
—, structure determination, 12
Tetrachloro-*o*-benzoquinone, reaction with styrene, 85
Tetracyanoethylene, reaction with butadiene, 4
Tetracyclone, reaction with triphenylcyclopropene, 21
—, see also Tetraphenylcyclopentadienone
Tetraenes, reaction with dienophiles, 18, 19
Tetrahydronaphthoquinones, formation of, 3
cis-1,2,5,6-Tetrahydrophthalic anhydride, formation of, 3
1,2,4,5-Tetramethylenecyclohexane, reaction with dienophiles, 17

1,2,3,4-Tetramethylnaphthalene, reaction with maleic anhydride, 14
Tetraphenylcyclopentadienone, reaction with maleic anhydride, 10
1,1,8,8-Tetraphenyloctatetraene, reaction with maleic anhydride, 18
Thermal association reactions, gaseous, 49
Thermal reactions, catalysis of, 65
Thiophen dioxides, reaction with dienophiles, 13
Thiophens, substituted, reaction with maleic anhydride, 11
Thiophyne, 23
Transition state, 74
—, non-polar, 90
—, stability of, 85, 89
Transoid conformation of dienes, 2, 89
Transoid dienes, reactivity of, 24
Triphenylcyclopropene, reaction with tetracyclone, 21
Triplets, cis-trans isomeric, of conjugated dienes, 72
—, of dienes, photo-sensitized production of, 71
Triptycene, formation of, 22
Tropone, reaction with maleic anhydride, 19
Two-step mechanism, of photosensitized reactions, 95
—, of uncatalysed thermal reactions, 93

Unsaturation, maximum accumulation of, 29

Van 't Hoff-Dimroth relation, 39
Vibrational modes, 45
Vinylarenes, reaction with dienophiles, 7
Vinyl compounds as dienophiles, 5
Vinylcycloalkenes, reaction with dienophiles, 7
4-Vinylcyclohexene, formation of, 72, 81, 94
Volume changes in condensed state reactions, 58